Mathematical Series 2

Lecture Notes on
ORDINARY DIFFERENTIAL EQUATIONS

by
REFAAT A. EL ATTAR
Professor of Mathematics
College of Engineering
Alexandria University
Egypt

Lulu Press Incorporated,
3131 RDU Center, Suite 210
Morrisville NC 27560, USA
http://www.lulu.com

ISBN: 1-4116-3920-0
ISBN13: 978-1-4116-3920-1
Printed in the United State of America

PREFACE

This book is written to provide an introduction to the subject of ordinary differential equation. The material presented can be covered in eight to ten 2-hour classroom lectures. Basic knowledge of calculus is needed. The book is written so that students and readers can use it as self study material. It is intended to help students in engineering and applied sciences understand various methods and techniques for solving differential equations.

I have collected many examples and problems on the subject that might help the reader getting on-hand experience with the techniques presented in this book. It is hoped that this work will give some motivation to the reader to dig a little bit further in the subject, especially the understanding of the existence of solutions and the rigorous theory of differential equations found in more advanced texts.

The text is divided into chapters. Chapter one deals with first order differential equations; several exercises and solved problems are given. In Chapter two, equations of first order and not of first degree are studied. Chapter three deals with second and higher order differential equations; in particular, linear equations with constant coefficients. Whereas, Chapter four treats linear equations with variable coefficients. Series solutions for ordinary differential equations are studied in Chapter five.

Finally, I express my sincere appreciation to my colleagues who used this note or part of it in their classrooms and/or supplied me with valuable comments and suggestions, especially Professor Mina B. Abdel Malek, Dr. Nagwa Badran and Dr. Mohamed Sayed.

The typing of the manuscript was a bit time consuming and mistakes might happen. I encourage the reader to point out to me the typing mistakes (or otherwise!) that I made.

REFAAT EL ATTAR

Alexandria, June 2, 2006

CONTENTS

Page

Page

Chapter One

Equations of First Order and First Degree

$$\frac{dy}{dx} = F(x, y)$$

Chapter 1.

Equations of First Order and First Degree

1.1. Introduction

By an ordinary differential equation we mean a relation between an independent variable x and an unspecified function y of x and certain of its derivatives $\frac{dy}{dx}, \frac{d^2y}{dx^2}, \ldots, \frac{d^ny}{dx^n}$ or $y', y'', \ldots, y^{(n)}$, where $', '', \ldots$ denote differentiation with respect to x. For example

$$\frac{dy}{dx} = 1 + y \tag{1}$$

$$y'' + 9y = 16\sin x \tag{2}$$

$$y'^2 - 2y'' + y - 3x = 0 \tag{3}$$

are ordinary differential equations.

For Equation (1), we say that $y = e^x - 1$ is a solution. However, this is not the only solution, for $y = 2e^x - 1$ also satisfies Equation (1). Therefore, an ordinary differential equation can have more than one solution depending on some conditions imposed on it.

The fundamental problem is to be sure that the differential equation has a solution or solutions and then to determine these solutions. This gives rise to the question of existence and uniqueness of the solution.

1.2. Basic Definitions and Terminology

We start the study of differential equations by giving some definitions and terminology. In general, a differential equation is an equation containing derivatives or differentials.

a. An ordinary differential equation of ***order*** n is an equation of the form

$$F(x, y, y', y'', \cdots, y^{(n)}) = 0 \tag{4}$$

For example

$$xy'' + 2y' + 3y - 6e^x = 0 \tag{5}$$

is an ordinary differential equation of order 2; while the equation

$$(y'')^2 - 2y\acute{y}'' + (y')^3 = 0 \tag{6}$$

is of order 3.

b. Equation (6) is a quadratic equation in the highest derivative y'''. We say that the equation is of ***degree*** 2.

c. An ordinary differential equation is said to be ***linear*** if it has the form

$$a_0(x)y^{(n)}+a_1(x)y^{(n-1)}+\cdots+a_{(n-1)}y'+a_n(x)y=Q(x) \qquad (7)$$

d. A solution (or particular solution) of the differential Equation (4) is a function $y=f(x)$, defined in some interval $a<x<b$ having derivatives up to the nth order throughout the interval such that Equation (4) becomes satisfied when y and its derivatives are replaced by $f(x)$ and its derivatives. For example, $y=e^{2x}$ is a solution of the equation

$$y''-4y=0 \qquad (8)$$

e. The differential equation may be expressed either in terms of derivatives or in terms of differentials. Thus

$$(2x+y^2)\frac{dy}{dx}-3x=1$$

and
$$(2x+y^2)dy-3x\,dx=dx$$

represent the same differential equation.

***Example* 1**: Show that $y=2x^3+c_1x+c_2$ is a solution of the differential equation $y''=12x$.

Solution: Since $y=2x^3+c_1x+c_2$ then we have $y'=6x^2+c_1$ and $y''=12x$.

Then $y=2x^3+c_1x+c_2$, which is called the ***primitive*** is a solution of the differential equation $y''=12x$.

***Example* 2**: Find the differential equation which has $y=c_1e^{-x}+c_2e^{x}+x$ as the general solution.

Solution: Differentiate the expression for y twice, we get

$y'=-c_1e^{-x}+c_2e^{x}+1$ and $y''=c_1e^{-x}+c_2e^{x}$.

Eliminating c_1 and c_2 from the above three equations, we obtain

$y''-y=-x$, which is the desired differential equation.

***Example* 3**: Find the differential equation which has

$y=c_1e^{3x}+c_2e^{-2x}+\sin x$ as the general solution.

Solution: Differentiate the expression for y twice, we get

$y'=3c_1e^{3x}-2c_2e^{-2x}+\cos x$ and

$y''=9c_1e^{3x}+4c_2e^{-2x}-\sin x$

In order to eliminate c_1 and c_2 from the above three equations, we set the determinant of the coefficients of c_1 and c_2 and the other terms equal to zero:

$$\begin{vmatrix} e^{3x} & e^{-2x} & y-\sin x \\ 3e^{3x} & -2e^{-2x} & y'-\cos x \\ 9e^{3x} & 4e^{-2x} & y''+\sin x \end{vmatrix} = 0$$

Simplifying, we finally obtain $y''-y'-6y+7\sin x+\cos x=0$, which is the desired differential equation. □

***Example* 4**: Find the differential equation representing the family of curves $y=c(x-c)^2$.

Solution: Differentiating the equation with respect to *x*, we get $y'=2c(x-c)$.

From the above two equations, we have

$$\frac{y'}{y}=\frac{2}{x-c} \quad \text{or} \quad c=x-\frac{2y}{y'}$$

Then, the required differential equation is $y'^3=4y(xy'-2y)$.

***Example* 5**: Show that $y=c_1x\cos(\ln x)+c_2x\sin(\ln x)+x\ln x$ is a solution of the differential equation $x^2y''-xy'+2y=x\ln x$.

Solution: Differentiating the given equation with respect to *x*, we get

$$y'=(c_2-c_1)\sin(\ln x)+(c_2+c_1)\cos(\ln x)+\ln x+1,$$

$$y''=-(c_2+c_1)\frac{\sin(\ln x)}{x}+(c_2-c_1)\frac{\cos(\ln x)}{x}+\frac{1}{x}$$

Multiplying the first equation by *x* and the second equation by x^2 and manipulating, we get $\quad x^2y''-xy'+2y=x\ln x$. □

Exercise 1.1

a. State the order and degree of the following differential equations:

1. $y''+3y'+2y=3x^3$
2. $x^2dy+2ydx=0$
3. $y'+y^2=x^2$
4. $D^2y=2x^3+3x-1$
5. $\left(\frac{d^3y}{dx^3}\right)^4-\left(\frac{d^2y}{dx^2}\right)^5+x=0$
6. $x^4\frac{dy}{dx}-x^3\frac{d^2y}{dx^2}=y^4\frac{d^3y}{dx^3}$
7. $\frac{d^3y}{dx^3}+\frac{d^2y}{dx^2}\cdot\frac{dy}{dx}+y=x$

b. Show that each of the following functions is a solution of the given differential equation:

1. $y = ax^2 - x$ $\qquad$ $xy' = 2y + x$
2. $y = c_1 \cos 2x + c_2 \sin 2x$ $\qquad$ $y'' + 4y = 0$
3. $y = c_1 e^{2x} + c_2 e^{-2x}$ $\qquad$ $y'' - 4y = 0$
4. $y = (2x + c)e^{-x}$ $\qquad$ $y' + y = 2e^{-x}$
5. $y = \ln \cos(x - a) + b$ $\qquad$ $y'' + (y')^2 + 1 = 0$
6. $y = c \sin^{-1} x$ $\qquad$ $y = y' \sin^{-1} x \sqrt{1 - x^2}$
7. $y = c_1 x + c_2 x^2$ $\qquad$ $x^2 y'' - 2xy' + 2y = 0$
8. $y = c_1 + 2x + c_2 x^2$ $\qquad$ $xy'' - y' + 2 = 0$
9. $x\, y = c_1 + c_2 x$ $\qquad$ $xy'' + 2y' = 0$
10. $y = c_1 e^{-2x} + c_2 e^{3x}$ $\qquad$ $y'' - y' - 6y = 0$
11. $y = c_1 \cos(2x + c_2)$ $\qquad$ $y'' + 4y = 0$

c. Find the differential equations for which the following functions are primitives:

1. $y = a/x + b$ $\qquad$ **Ans:** $y'' + 2y'/x = 0$
2. $y = a \cos 3x + \sin 2x$ $\qquad$ **Ans:** $\cos 2x \cdot y' + 2 \sin 2x \cdot y = 2$
3. $y = a \ln x$ $\qquad$ **Ans:** $y' = y/(x \ln x)$
4. $x^2 - y^2 = 1$ $\qquad$ **Ans:** $y\, y' - x = 0$
5. $y = ae^{3x} + bx\, e^{3x}$ $\qquad$ **Ans:** $y'' - 6y' + 9y = 0$
6. $y = \dfrac{a + x}{1 - ax}$ $\qquad$ **Ans:** $y' = \dfrac{1 + y^2}{1 + x^2}$
7. $y = \cos 2x + 2 \cos 3x + 3 \sin 3x$ $\qquad$ **Ans:** $y'' + 9y = 5 \cos 2x$

d. Find the differential equation representing all circles with radius *a*.

Ans: $(y'^2 + 1)^3 = a^2 y''^2$

e. Find the differential equation representing all circles passing through the origin and having their centers on the *x*-axis. **Ans:** $2xy' = y^2 - x^2$

1.3. Equations of First Order and First Degree

The general form of a first order and first degree differential equations is

$$\frac{dy}{dx} = F(x, y) \tag{9}$$

The solution of this equation depends on the function $F(x, y)$. In our previous discussion, we have assumed that for each differential equation there is a general solution. The following theorem gives the conditions for the ***existence*** and ***uniqueness*** of such solution. The proof of this theorem reaches into other branches of mathematics and is omitted here.

Theorem: If $F(x, y)$ is continuous and has a continuous partial derivative with respect to y at each point of the region R defined by $|x - x_0| < \delta$, $|y - y_0| < \delta$, then there exists one and only one solution (a unique solution) of Equation (9), in this region R, that passes through the point (x_0, y_0).

The restrictions imposed on $F(x, y)$ are ***sufficient*** conditions. They are not ***necessary*** conditions, i.e., it is possible that a solution exists although these conditions are not satisfied. For example, in the differential equation

$$\frac{dy}{dx} = -\frac{x}{y},$$

the function $F(x, y) = -x/y$ and its partial derivative with respect to y exist everywhere, except at $y = 0$, where neither is defined. However, it can be shown that $x^2 + y^2 = c^2$, where c is an arbitrary constant, satisfies the differential equation.

For the problem of uniqueness, consider the differential equation

$$\frac{dy}{dx} = \frac{x}{y}.$$

It can be shown that the solution is $y = cx$, which is a family of lines passing through the origin. Then at any point (x_0, y_0) except $(0, 0)$ a unique line will pass. However, all the lines in the family, not a unique one, will pass through the origin.

There is no general method to solve first order differential equations, but we can find different techniques for particular cases depending on the form of the equation. We shall give some of these techniques.

1.3.1. Separation of Variables

If $F(x, y)$ in Equation (9) takes the form $\dfrac{f(x)}{g(y)}$, $g(y) \neq 0$, where $f(x)$ is a function of x alone and $g(y)$ is a function of y alone, then we can write

$$g(y)\,dy = f(x)\,dx \tag{10}$$

and integrating Equation (10) gives

$$\int g(y)\,dy = \int f(x)\,dx + c \tag{11}$$

where c is an arbitrary constant.

***Example* 1**: Find the general solution of $\dfrac{dy}{dx} = -\dfrac{x}{y}$.

Solution: Rearranging we get $x\,dx + y\,dy = 0$.

Integrating we get $x^2 + y^2 = c^2$, $c > 0$;

this is the general solution of the differential equation. □

***Example* 2**: Find the general solution of $y\,dx - x\,dy = 0$, $y \neq 0$ and $x \neq 0$.

Solution: Rearranging, we get $\dfrac{dy}{y} - \dfrac{dx}{x} = 0$. Integrating, we get

$\ln y - \ln x = c$, $x > 0$, $y > 0$ or $\ln \dfrac{x}{y} = c$

or $y = kx$ where $k = e^c$. □

***Example* 3**: Find the general solution of $\dfrac{dy}{dx} = \cos^2 y \cos x$.

Solution: Separating the variables and integrating, we get

$\int \sec^2 y\,dy = \int \cos x\,dx$ or $\tan y = \sin x + c$. □

***Example* 4**: Solve the equation $x^2(1+y)\dfrac{dy}{dx} + (1-x)y^2 = 0$.

Solution: Separating the variables, we get $\dfrac{1+y}{y^2}dy = -\dfrac{1-x}{x^2}dx$

Integrating, we get

$$\int \left(\frac{1}{y^2} + \frac{1}{y}\right) dy = \int \left(-\frac{1}{x^2} + \frac{1}{x}\right) dx,$$

Hence $\ln \dfrac{y}{x} - \dfrac{1}{x} - \dfrac{1}{y} = c$. □

***Example* 5**: Solve the equation $(x^2+1)(y^2-1)dx+xy\,dy=0$.

Solution: Separating the variables, we get $\dfrac{x^2+1}{x}dx=-\dfrac{y}{y^2-1}dy$.

Integrating, we get $x^2+\ln x^2+\ln(y^2-1)=\ln c$.

Hence $y^2=1+\dfrac{ce^{-x^2}}{x^2}$. □

***Example* 6**: Find the general solution of the following differential equation as well as the particular curve that passes through the point $(0,0)$:

$e^x\cos y\,dx+(1+e^x)\sin y\,dy=0$

Solution: Separating the variables, we obtain $\tan y\,dy=-\dfrac{e^x}{1+e^x}dx$.

Integrating, we get $\ln\cos y+\ln c=\ln(1+e^x)$ or $c\cos y=1+e^x$.

This is the general solution. To obtain the particular curve that passes through the point $(0,0)$, we substitute this point into the general solution to get $1+1=(c)(1)$ or $c=2$.

Then the particular curve is $1+e^x=2\cos y$. □

Note: Some differential equations take the form $\dfrac{dy}{dx}=f(ax+by+c)$ and are not separable. In this case a transformation of the form $v=ax+by+c$ will make them separable.

***Example* 7**: Solve the equation $\dfrac{dy}{dx}=(4x+y+1)^2$.

Solution: Let $v=4x+y+1$, then $\dfrac{dv}{dx}=4+\dfrac{dy}{dx}$ or $\dfrac{dy}{dx}=\dfrac{dv}{dx}-4$, and the equation becomes $\dfrac{dv}{dx}-4=v^2$ or $\dfrac{dv}{4+v^2}=dx$.

Integrating, we get $\dfrac{1}{2}\tan^{-1}\dfrac{v}{2}=x+c$ or $v=2\tan(2x+2c)$,

and the solution is $4x+y+1=2\tan(2x+2c)$. □

Exercise 1.2

a. Find all solutions by separation of variables:

1. $y' = e^{x+y}$ **Ans:** $e^{x} + e^{-y} = c$

2. $y' = e^{x-y} + x^{2}e^{-y}$ **Ans:** $e^{y} = e^{x} + \frac{1}{3}x^{3} + c$

3. $y' = x^{3}y^{-2}$ **Ans:** $4y^{3} = 3x^{4} + c, y \neq 0$

4. $y' = (y-1)(y-2)$ **Ans:** $y - 2 = ce^{x}(y-1)$

5. $y' = 3y$ **Ans:** $y = ce^{3x}$

6. $\sin x \cos y\, dx + \tan y \cos x\, dy = 0$ **Ans:** $\cos x\, e^{-\sec y} = c$

7. $\cot x\, dx + [e^{y}/(e^{y}+1)]dy = 0$ **Ans:** $(\sin x)(e^{y}+1) = c$

8. $e^{x-y}dx + e^{y-x}dy = 0$ **Ans:** $e^{2x} + e^{2y} = c$

9. $y' = xy^{2} + y^{2} + xy + y$ **Ans:** $\ln\frac{y}{y+1} = \frac{1}{2}x^{2} + x + c$

10. $y' = e^{x+y}\cosh x$ **Ans:** $e^{-y} + \frac{1}{4}e^{2x} + \frac{1}{2}x = c$

11. $\frac{dy}{dx} = \tan^{2}(x+y)$ **Ans:** $2(x - y + 2c) = \sin[2(x+y)]$

12. $\frac{dy}{dx} = \frac{\sin x + x\cos x}{y(2\ln y + 1)}$ **Ans:** $x\sin x = y^{2}\ln y + c$

13. $\sin x\frac{dy}{dx} + y\cos x = 0$ **Ans:** $y\sin x = c$

14. $(4x + xy^{2})dx + (y + x^{2}y)dy = 0$ **Ans:** $(1+x^{2})(4+y^{2}) = c$

15. $(xy + x)dy - (xy + y)dx = 0$ **Ans:** $x = cye^{y-x}$

16. $(1+x)y\, dx + (1-y)x\, dy = 0$ **Ans:** $xy = ce^{y-x}$

17. $y\, dx + (1+x^{2})\tan^{-1}x\, dy = 0$ **Ans:** $y\tan^{-1}x = c$

18. $(1+x^{2})dy + x\sqrt{1-y^{2}}\, dx = 0$ **Ans:** $2\sin^{-1}y + \ln(1+x^{2}) = c$

19. $y' = \sec y\tan x$ **Ans:** $\ln\cos x + \sin y = c$

20. $(x+y)^2 y' = 1$ **Ans:** $y - \tan^{-1}(x+y) = c$

21. $y' = \sec(x+y)$ **Ans:** $y - \tan[(x+y)/2] = c$

22. $y' = \sin(x+y) + \cos(x+y)$ **Ans:** $x + c = \ln\{1 + \tan[\{x+y)/2]\}$

23. $y' + 1 = e^{x+y}$ **Ans:** $x + e^{-(x+y)} = c$

24. $(x+y+1)\dfrac{dy}{dx} = 1$ **Ans:** $x + y + 2 = ce^{y}$

25. $y' = e^{2x+y-1} - 2$ **Ans:** $y = 1 - 2x - \ln(c - x)$

26. $(x+y)^2 \dfrac{dy}{dx} = 4$ **Ans:** $y - 2\tan^{-1}[(x+y)/2] = c$

27. $y' = (x-y+1)^2 + x - y$ **Ans:** $y = 1 + x + \dfrac{1+2ce^{-3x}}{1-ce^{-3x}}$

28. $\sin\theta\, d\rho + \rho\cos\theta\, d\theta = 0$ **Ans:** $\rho\sin\theta = c$

29. $d\rho + \rho\tan\theta\, d\theta = 0$ **Ans:** $\rho = c\cos\theta$

b. Find the particular solutions satisfying the given conditions for the following differential equations:

1. $x\,dx + y\,dy = 0;\ y = 2$ when $x = 1$ **Ans:** $x^2 + y^2 = 5$

2. $dy = y\tan x\, dx;\ \ y = 1$ when $x = 0$ **Ans:** $y = \sec x$

3. $e^x \sec y\, dx + (1+e^x)\sec y \tan y\, dy = 0;\ \ y = \pi/3$ when $x = 3$

Ans: $1 + e^x = 2(1+e^3)\cos y$

4. $y' + y^2\sin x = 0;\ y = 0$ when $x = 0$ **Ans:** $y = 0$

5. $y' + 2y = 0;\ y = 100$ when $x = 0$ **Ans:** $y = 100e^{-2x}$

6. $dy = x(2y\,dx - x\,dy);\ y = 4$ when $x = 1$ **Ans:** $y = 2 + 2x^2$

c. Show that the substitution $v = ax + by + c$ will transform the equation $y' = f(ax+by+c)$ into an equation that can be solved by separation of variables.

d. Find a solution of the equation $(1-x^2)dy + 4xy\,dx = 0$ that satisfies that $y = 9$ when $x = -2$, $y = 2$ when $x = 0$ and $y = 0$ when $x = 2$.

e. Find the equation of the curve that passes through the point (1, 2) and has a slope at any point of $y' = -\dfrac{2xy}{x^2+1}$. **Ans:** $y(1+x^2) = 4$

1.3.2. Homogeneous Equations

A first order differential equation is said to be ***homogeneous*** if it can be put in the form

$$\frac{dy}{dx}=f\left(\frac{y}{x}\right) \tag{12}$$

The following equations are of this type

$$y'=\frac{y}{x},\quad y'=\frac{x^2-y^2}{xy},\quad y'=\sin\frac{y}{x}$$

Using the substitution $y/x=v$ or $y=vx$, where v is a function of x, the equation reduces to $v+x\dfrac{dv}{dx}=f(v)$ or $x\,dv+(f(v)-v)dx=0$, which is one that can be solved by separation of variables. And the solution is

$$\ln x=\int\frac{dv}{f(v)-v}+c.$$

The following examples illustrate the procedure.

***Example* 1**: Solve the equation $y'=\dfrac{x^2+y^2}{xy}$.

Solution: Let $y=vx$, then $\dfrac{dy}{dx}=v+x\dfrac{dv}{dx}$, therefore

$$v+x\frac{dv}{dx}=\frac{x^2(1+v^2)}{x^2v}.$$

Rearranging, we get $v\,dv-\dfrac{dx}{x}=0$.

Integrating, we get

$$\frac{1}{2}v^2-\ln x=\frac{1}{2}c,\ \text{ or }\ y^2=x^2\ln x^2+cx^2.$$

□

***Example* 2**: Find the general solution of $(x^4-2xy^3)\dfrac{dy}{dx}=2x^3y-y^4$.

Solution: Rearranging, we get $\dfrac{dy}{dx}=\dfrac{2x^3y-y^4}{x^4-2xy^3}$, which is a homogeneous equation. Let $y=vx$, then $v+x\dfrac{dv}{dx}=\dfrac{2v-v^4}{1-2v^3}$.

Separating the variables and integrating, we get

$$\int \frac{dx}{x} = \int \frac{(1-2v^3)dv}{v(1+v)(1-v+v^2)}.$$

Using partial fraction, we finally obtain

$$\ln cx = \ln v - \ln(1+v) - \ln(1-v+v^2) = \ln\left\{\frac{v}{1+v^3}\right\}.$$

Back-substituting, the general solution is $c(x^3+y^3)=xy$. □

***Example* 3**: Find the general solution of $x\,dy - y\,dx = \sqrt{x^2+y^2}\,dx$.

Solution: Rearranging, we get $\dfrac{dy}{dx} = \dfrac{y+\sqrt{x^2+y^2}}{x}$, which is clearly a homogeneous equation.

Let $y = vx$, then $v + x\dfrac{dv}{dx} = v + \sqrt{1+v^2}$.

Rearranging and integrating, we get $\displaystyle\int \frac{dx}{x} = \int \frac{dv}{\sqrt{1+v^2}}$, or

$\ln x + \ln c = \ln(v+\sqrt{1+v^2})$, or $cx = v+\sqrt{1+v^2}$.

Back-substituting, the general solution is $cx^2 = y+\sqrt{x^2+y^2}$. □

Note: In some instance, if we use of the substitution $y = vx$, we may encounter integrals that are difficult to evaluate in closed form. In this case one might use the reverse substitution $x = uy$ instead, see the exercises below.

Exercise 1.3

a. Verify that the following equations are homogeneous and find all solutions:

1. $y' = \dfrac{x-y}{x+y}$ **Ans**: $x^2 - 2xy - y^2 = c, \quad x+y \neq 0$
2. $x\,y' - y = x\,e^{y/x}$ **Ans**: $e^{-y/x} + \ln x = c, \quad x \neq 0$
3. $(2x^3+y^3)dx - 3xy^2 dy = 0$ **Ans**: $y^3 - x^3 = cx$
4. $(3x^2y+y^3)dx + (x^3+3xy^2)dy = 0$ **Ans**: $x^3y + xy^3 = c$
5. $y' = \dfrac{y}{x} + \sin\dfrac{y-x}{x}$ **Ans**: $\operatorname{cosec}\dfrac{y-x}{x} - \cot\dfrac{y-x}{x} = cx, \quad x \neq 0$

6 $x(1+e^{y/x})dy + e^{y/x}(x-y)dx = 0$ **Ans**: $y + xe^{y/x} = c$

7. $y' = \frac{y}{x} + \sin\frac{y}{x}$ **Ans**: $\tan\frac{y}{2x} = cx$,

8. $(y+x)dy + \frac{y^2}{x}dx = 0$ **Ans**: $y^2x = c(2y+x)$

9. $x\frac{dy}{dx} - y = \sqrt{x^2-y^2}$ **Ans**: $\sin^{-1}\frac{y}{x} - \ln x = c$

10. $x\,y\,dy + x\sqrt{x^2+y^2}dx = 0$ **Ans**: $x = c\,e^{-(1/x)\sqrt{x^2+y^2}}$

11. $\frac{dy}{dx} = \frac{5x+4y}{2x-y}$ **Ans**: $\ln(y^2+5x^2+2xy) - 3\tan^{-1}\left(\frac{y+x}{2x}\right) = c$

12. $(2\sqrt{xy} - y)dx - x\,dy = 0$ **Ans**: $\sqrt{xy} - x = c$

13. $2x^3y\,dx + (x^4+y^4)dy = 0$ **Ans**: $3x^4y^2 + y^6 = c$

14. $x\cos\left(\frac{y}{x}\right)(y\,dx + x\,dy) = y\sin\left(\frac{y}{x}\right)(x\,dy - y\,dx)$

Ans: $x\,y\cos(y/x) = c$

b. Find the solution that satisfies the given initial conditions:

1. $(xy^2 - x^2y)dx - x^3dy = 0$, when $x=1$, $y=1$ **Ans**: $y = 2x/(x^2+1)$

2. $\frac{dx}{dt} = \frac{xt}{x^2+t^2}$, when $t=0, x=1$ **Ans**: $x = e^{t^2/2x^2}$

3. $x\frac{dy}{dx} = y + \sqrt{4x^2+y^2}$, when $x=1$, $y=0$ **Ans**: $2x^2 = y + \sqrt{4x^2+y^2}$

4. $x\,y^2\frac{dy}{dx} = y^3 - x^3$, when $x=1$, $y=2$ **Ans**: $y^3 + 3x^3\ln x = 8x^3$

5. $(x+\sqrt{x\,y})\frac{dy}{dx} + x - y = y\sqrt{y/x}$, when $x=1$, $y=1$

Ans: $3x\sqrt{x}\ln x + 3\sqrt{x}\,y + 2y\sqrt{y} = 5x\sqrt{x}$

c. If $u(x,y)dx + v(x,y)dy = 0$ is a homogeneous equation, show that the substitution $x = r\cos\theta$, $y = r\sin\theta$ will reduce the equation to one that can be solved by separation of variables.

1.3.3. Equations of the Form $\dfrac{dy}{dx}=\dfrac{ax+by+c}{a_1x+b_1y+c_1}$

Clearly, equations of this form are not homogeneous, but using the substitution $x = X+h$, $y = Y+k$, they reduce to homogeneous equations; where h and k are chosen such that

$$\left.\begin{aligned} ah+bk+c&=0 \\ a_1h+b_1k+c_1&=0 \end{aligned}\right\} \qquad (13)$$

This method will fail if $\dfrac{a_1}{a}=\dfrac{b_1}{b}$. In this particular case, a substitution of the form $v=ax+by$ will reduce the equation to one that can be solved by separation of variables. The following examples illustrate the procedure.

***Example* 1**: Solve the differential equation $\dfrac{dy}{dx}=\dfrac{x-y}{x+1}$.

Solution: Let $x = X+h$ and $y = Y+k$. Choose h and k such that $h-k=0$, $h+1=0$, then $h = -1$, $k = -1$. The differential equation becomes $\dfrac{dY}{dX}=\dfrac{X-Y}{X}$. This is a homogeneous equation.

Let $Y = vX$, then $v+X\dfrac{dv}{dX}=1-v$.

Separating the variables and integrating, we obtain

$$\int\frac{dX}{X}=\int\frac{dv}{1-2v}\text{, i.e., } \ln cX=-\frac{1}{2}\ln(1-2v)$$

or $\quad X^2(1-2\dfrac{Y}{X})=c$. Back-substituting $X = x+1$ and $Y = y+1$, we get $(x+1)(x-2y-1)=c$. □

***Example* 2**: Solve the equation $\dfrac{dy}{dx}=\dfrac{9x+12y-3}{3x+4y+2}$.

Solution: We can see that $\dfrac{a_1}{a}=\dfrac{b_1}{b}=3$. Then let $v = 3x+4y$, then

$\dfrac{dv}{dx}=3+4\dfrac{dy}{dx}$. Hence $\dfrac{dy}{dx}=\dfrac{1}{4}\dfrac{dv}{dx}-\dfrac{3}{4}$, the differential equation becomes $\dfrac{1}{4}\dfrac{dv}{dx}-\dfrac{3}{4}=\dfrac{3v-3}{v+2}$, or $\dfrac{dv}{dx}=\dfrac{5v-6}{v+2}$.

Separating the variables and integrating and back-substituting for v, we get $5y-15x+3\ln(15x+20y-2)=c$. □

Exercises 1.4

a. Solve the following differential equations:

1. $\dfrac{dy}{dx}=\dfrac{3x-y+3}{3x-y-1}$ **Ans**: $2\ln(3x-y-3)=y-x+c$

2. $\dfrac{dy}{dx}=\dfrac{x+y+3}{x-y-5}$ **Ans**: $\ln[c(x-1)]=\tan^{-1}\dfrac{y+4}{x-1}-\dfrac{1}{2}\ln\dfrac{(x+1)^2+(y+4)^2}{(x-1)^2}$

3. $\dfrac{dy}{dx}=\dfrac{2x+y-2}{x-y-2}$ **Ans**: $2x^2+2x(y-2)-(y-2)^2=c$

4. $\dfrac{dy}{dx}=\dfrac{x-1}{3x-2y-5}$ **Ans**: $(2y-x+3)^2=c(y-x+2)$

5. $\dfrac{dy}{dx}=\dfrac{y-2x}{4x+y-6}$ **Ans**: $(x+y-3)^3=c(2x+y-4)^2$

6. $\dfrac{dy}{dx}=\dfrac{-2}{2x-y+3}$ **Ans**: $y+c=-\ln(2x-y+4)$

7. $\dfrac{dy}{dx}=\dfrac{x+y-1}{x+y+1}$ **Ans**: $x-y+c=\ln(x+y)$

8. $\dfrac{dy}{dx}=\dfrac{x+y+1}{2x+2y+3}$ **Ans**: $x+y+\dfrac{4}{3}=ce^{3(x-2y)}$

9. $\dfrac{dy}{dx}=\dfrac{2y+x-3}{2x+y-3}$ **Ans**: $c(x-y)^2=x+y-2$

10. $\dfrac{dy}{dx}=\dfrac{6y+4x+5}{2x+3y+4}$ **Ans**: $y-2x+\frac{3}{8}\ln(16x+24y+23)=c$

11. $(x-y)\,dy=(x+y+1)\,dx$

    Ans: $\tan^{-1}[(2y+1)/(2x+1)]=\ln(c\sqrt{x^2+y^2+x+y+1/2})$

b. Solve the following initial value problems:

1. $\dfrac{dy}{dx}=\dfrac{x+2y+1}{2x+4y-1}$; when $x=0,\ y=0$

    Ans: $4x-8y+3\ln(4x+8y+1)=0$

2. $\dfrac{dy}{dx}=\dfrac{2x+y+7}{6x+3y}$; when $x=1,\ y=-2$

    Ans: $3y-x+7=3\ln(2x+y+1)$

1.3.4. Exact Equations

The differential equation

$$u(x, y)\,dx + v(x, y)\,dy = 0 \tag{14}$$

is said to be ***exact*** if it can be put in the differential form $dz = 0$, where $z = f(x, y)$. The solution is then $z =$ *constant*.

From the analysis of functions of several variables, we have

$$dz = \frac{\partial z}{\partial x}dx + \frac{\partial z}{\partial y}dy \tag{15}$$

Comparing Equation (14) with Equation (15), we have

$$\frac{\partial z}{\partial x} = u\,, \quad \frac{\partial z}{\partial y} = v\,, \quad \text{also} \quad \frac{\partial^2 z}{\partial x\,\partial y} = \frac{\partial u}{\partial y}\,, \quad \frac{\partial^2 z}{\partial y\,\partial x} = \frac{\partial v}{\partial x}\,.$$

And if z and its first and second partial derivatives are continuous in some interval then $\dfrac{\partial^2 z}{\partial x\,\partial y} = \dfrac{\partial^2 z}{\partial y\,\partial x}$, therefore

$$\frac{\partial u}{\partial y} = \frac{\partial v}{\partial x} \tag{16}$$

Equation (16) represents a necessary condition for Equation (14) to be exact.

To show that the condition is also sufficient, we proceed as follows. Assume that the condition of Equation (16) hold, then we will show that Equation (14) is exact. Let $g(x, y) = \int u\,dx$ be the partial integral of u, i.e., the integral obtained by keeping y fixed, then

$$\frac{\partial}{\partial x}\left(v - \frac{\partial g}{\partial y}\right) = \frac{\partial v}{\partial x} - \frac{\partial^2 g}{\partial x\,\partial y} = \frac{\partial v}{\partial x} - \frac{\partial^2 g}{\partial y\,\partial x}$$
$$= \frac{\partial v}{\partial x} - \frac{\partial}{\partial y}\left(\frac{\partial g}{\partial x}\right) = \frac{\partial v}{\partial x} - \frac{\partial u}{\partial y} = 0$$

Now,
$$f(x, y) = g(x, y) + \int\left(v - \frac{\partial g}{\partial y}\right)dy$$

Then the total differential of $f(x, y)$ is

$$df = dg + \left(v - \frac{\partial g}{\partial y}\right)dy = \left(\frac{\partial g}{\partial x}dx + \frac{\partial g}{\partial y}dy\right) + v\,dy - \frac{\partial g}{\partial y}dy$$
$$= \frac{\partial g}{\partial x}dx + v\,dy = u\,dx + v\,dy$$

Therefore, if Equation (16) is satisfied, Equation (14) is an exact equation.

Remember that the solution is now $z = f(x, y) = c$. To summarize, the following theorem is a test for an exact differential equation.

Theorem: If $u(x,y)$ and $v(x,y)$ are continuous and have continuous first partial derivatives is the region R defined by $a<x<b$, $c<y<d$, then a necessary and sufficient condition that the differential equation (14) be exact is $\frac{\partial u}{\partial y}=\frac{\partial v}{\partial x}$.

We give here examples to illustrate the procedure to solve exact equations.

***Example* 1**: Solve the equation

$$(3x^2y+2xy)\,dx+(x^3+x^2+2y)\,dy=0.$$

Solution: We have $u=3x^2y+2xy$ and $v=x^3+x^2+2y$, then

$$\frac{\partial u}{\partial y}=3x^2+2x \text{ and } \frac{\partial v}{\partial x}=3x^2+2x,$$

then the equation is exact. Hence

$$\frac{\partial z}{\partial x}=3x^2y+2xy,\quad \frac{\partial z}{\partial y}=x^3+x^2+2y$$

Integrating, we get

$$z=x^3y+x^2y+\phi(y),\quad z=x^3y+x^2y+y^2+\psi(x).$$

Comparing these two values of z, we conclude that

$$\phi(y)=y^2 \text{ and } \psi(x)=0.$$

Then the solution is $x^3y+x^2y+y^2=c$. □

Note: This solution could have been obtained directly for exact equations by following these steps:

1. Integrate u with respect to x, and
2. Add the integration of v with respect to y (the terms not containing x).
3. Equate the result to an arbitrary constant.

***Example* 2**: Solve the equation

$$(xy\cos xy+\sin xy)\,dx+(x^2\cos xy+e^y)\,dy=0.$$

Solution: We have $u=xy\cos xy+\sin xy$ and $v=x^2\cos xy+e^y$,

Then $\frac{\partial u}{\partial y}=2x\cos xy-x^2y\sin xy=\frac{\partial v}{\partial x}$.

Then the equation is exact, and the solution is

$$\int (xy\cos xy + \sin xy)\,dx + \int e^y\,dy = c$$

Hence $x\sin xy + e^y = c$. □

***Example* 3**: Solve the differential equation:

$$(3yx^2 - y/x^2)\,dx + (x^3 + \cos y + 1/x)\,dy = 0.$$

Solution: The equation is exact because $\dfrac{\partial u}{\partial y} = 3x^2 - \dfrac{1}{x^2} = \dfrac{\partial v}{\partial x}$.

Then the solution is $x^3 y + \dfrac{y}{x} + \sin y = c$. □

Note: If the equation $u(x,y)\,dx + v(x,y)\,dy = 0$ is not exact, it can be made exact by multiplying by a function $\mu(x,y)$. This function is called the Integrating factor. If this integrating factor is obtained then the said condition for exactness becomes $\dfrac{\partial}{\partial y}(\mu u) = \dfrac{\partial}{\partial x}(\mu v)$. We now give the following theorem.

Theorem: The differential equation $u(x,y)\,dx + v(x,y)\,dy = 0$ possesses an infinite number of integrating factor.

Proof: Let $\mu(x,y)$ be an integrating factor, then by definition

$\mu\left[u(x,y)\,dx + v(x,y)dy\right] = 0$ must be exact.

Therefore, there must exist a function $z(x,y)$ such that

$dz = \mu\left[u(x,y)\,dx + v(x,y)dy\right] = 0$.

This implies that $z(x,y) = \text{constant}$, which is a solution of the differential equation.

Now, assume that $g(z)$ is any function of z, then we have

$g(z)dz = \mu\, g(z)\left[u(x,y)\,dx + v(x,y)dy\right]$

Since the expression to the left of this equation is an exact differential, the expression to the right must also be an exact differential. It follows that $\mu\, g(z)$ is an integrating factor of the differential equation. And, since $g(z)$ is an arbitrary function of z, it follows that the differential equation possesses an infinite number of integrating factors.

***Example* 4**: Show that the equation $(3xy^2 + 2y)\,dx + (2x^2 y + x)\,dy = 0$ is not exact but it becomes exact by multiplying by *x*, and find its solution.

Solution: We have, $u = 3xy^2 + 2y$ and $v = 2x^2y + x$, then

$$\frac{\partial u}{\partial y} = 6xy + 2 \text{ and } \frac{\partial v}{\partial x} = 4xy + 1,$$

then the equation is not exact. Multiplying by x the equation becomes

$(3x^2y^2 + 2xy)\,dx + (2x^3y + x^2)\,dy = 0$,

Now, for this equation we have

$u = 3x^2y^2 + 2xy$ and $v = 2x^3y + x^2$.

Then $\dfrac{\partial u}{\partial y} = 6x^2y + 2x = \dfrac{\partial v}{\partial x}$.

The equation is exact, and the solution is $x^2y + x^3y^2 = c$ □

Exercise 1.5

a. Show that the differential equations are exact and find all solutions:

1. $y \sin 2x\, dx - (y^2 + \cos^2 x)\,dy = 0$ **Ans:** $y \cos 2x + \frac{2}{3}y^3 + y = c$

2. $[y(1 + 1/x) + \cos y]\,dx + (x + \ln x - x \sin y)\,dy = 0$

   Ans: $xy + y \ln x + x \cos y = c$

3. $y' = (2x - y)/(x + 2y - 5)$ **Ans:** $x^2 - xy + y^2 + 5y = c$

4. $(e^y + 1)\cos x\, dx + e^y \sin x\, dy = 0$ **Ans:** $(e^y + 1)\sin x = c$

5. $\sec^2 x \tan y\, dx + \sec^2 y \tan x\, dy = 0$ **Ans:** $\tan y \tan x = c$

6. $(y \cos xy + e^x)\,dx + (x \cos xy - 2ye^{y^2})\,dy = 0$

   Ans: $\sin xy + e^x - e^{y^2} = c$

7. $y \sin 2x\, dx - (y^2 + \cos^2 x)\,dy = 0$ **Ans:** $y \cos 2x + \frac{2}{3}y^3 + y = c$

8. $2xy\, dx + (x^2 + 1)\,dy = 0$ **Ans:** $x^2y + y = c$

9. $[y(1 + 1/x) + \cos y]\,dx + (x + \ln x - x \sin y)\,dy = 0$

   Ans: $xy + y \ln x + x \cos y = c$

10. $(1 + e^{x/y})\,dx + e^{x/y}[1 - (x/y)]\,dy = 0$ **Ans:** $x + e^{x/y} = c$

11. $(2ye^{2x} - y \cos xy + 2x)\,dx + (e^{2x} - x \cos xy)\,dy = 0$

    Ans: $ye^{2x} - \sin xy + x^2 + c = 0$

12. $(1+\ln xy)\,dx+(1+x/y)\,dy=0,\ \ x>0, y>0$ **Ans:** $y+x\ln xy=c$

13. $(2x+\cosh xy)\,dx+[(xy\cosh xy-\sinh xy)/y^2]\,dy=0$

Ans: $x^2y+\sinh xy=cy$

14. $dy=\left(\dfrac{y}{x}-\operatorname{cosec}^2\dfrac{y}{x}\right)dx$ **Ans:** $2y-x\sin\dfrac{2y}{x}+4x\ln x=cx$

b. Show that the following equations are not exact, then use the given integrating factor to make them exact and find all solutions.

1. $(xy\sin xy+\cos xy)y\,dx+(xy\sin xy-\cos xy)x\,dy=0$

   $[\mu=1/(2xy\cos xy)]$ **Ans:** $(x\sec xy)/y=c$

2. $(x^2+y^2+2x)\,dx+2y\,dy=0$ $[\mu=e^x]$ **Ans:** $e^x(x^2+y^2)=c$

c. Find a solution of each of the following differential equations that satisfies the given initial condition:

1. $x^2dx+ye^y\,dy=0$; when $x=0, y=1$ **Ans:** $x^3+3e^y(y-1)=0$

2. $e^{x^2}dx+\sin(1+y^2)\,dy=0$; when $x=0, y=0$

   Ans: $\int_0^x e^{t^2}dt+\int_0^y \sin(1+t^2)\,dt=0$

3. $(3x+y+5)^3(3dx+dy)+2\,dx-dy=0; x=-2,\ y=1$

   Ans: $(3x+y+5)^4+4(2x-y+5)=0$

4. $(x+y)\,dx+(x+2y)\,dy=0$; when $x=2, y=3$

   Ans: $x^2+2xy+2y^2=34$

5. $\tan y+\dfrac{y}{1+x^2}=(2\tan^{-1}y-\tan^{-1}x-x\sec^2 y)\,y'$; when $x=0, y=1$

   Ans: $x\tan y+y\tan^{-1}x-2y\tan^{-1}y+\ln(1+y^2)+\dfrac{\pi}{2}=\ln 2$

1.3.5. Linear Equations

A linear first order ordinary differential equation takes the form

$$\frac{dy}{dx}+P(x)y=Q(x) \tag{17}$$

Examples are $\frac{dy}{dx}+4xy=\cos x\ ;\ \frac{dy}{dx}+\frac{2}{x}y=8x\ .$

To solve these equations, suppose we multiply Equation (17) by an arbitrary function μ(x), then

$$\mu\frac{dy}{dx}+\mu Py=\mu Q \tag{18}$$

or

$$\frac{d}{dx}(\mu y)-\left(\frac{d\mu}{dx}-\mu P\right)y=\mu Q \tag{19}$$

Now, since μ(x) is arbitrary, let us choose it such that

$$\frac{d\mu}{dx}-\mu P=0 \tag{20}$$

Hence

$$\mu=e^{\int P\,dx} \tag{21}$$

μ(x) is said to be the ***integrating factor*** for the differential equation. Note that we need not use a constant of integration since this will not affect the result. Equation (17) reduces to

$$\frac{d}{dx}(\mu y)=\mu Q$$

or

$$y=\frac{1}{\mu}\left\{\int\mu Q\,dx+c\right\} \tag{22}$$

is the solution.

Note: The procedure is now:

1. Evaluate the integrating factor μ from $\mu=e^{\int P\,dx}$, and

2. Write the solution y as $y=\frac{1}{\mu}\left\{\int\mu Q\,dx+c\right\}$.

***Example* 1**: Solve the differential equation $\frac{dy}{dx}+2xy=8x\ .$

Solution: The integrating factor is $\mu=e^{\int 2x dx}=e^{x^2}$. The solution will be

$$y=e^{-x^2}\left[\int e^{x^2}\cdot 8x\,dx+c\right]=e^{-x^2}\left[4\int e^{x^2}dx^2+c\right]=4+c\,e^{-x^2}\ \square$$

***Example* 2**: Solve the equation $\frac{dy}{dx}+(4\cot x)\,y=\cot x\ \operatorname{cosec} x$.

Solution: The integrating factor is $\mu=e^{\int 4\cot x\,dx}=e^{4\ln\sin x}=\sin^4 x$. The solution is

$$y=\frac{1}{\sin^4 x}\left\{\int \sin^4 x\cdot\frac{\cos x}{\sin^2 x}dx+c\right\}=\frac{1}{\sin^4 x}\left\{\frac{1}{3}\sin^3 x+c\right\}$$

$$=\frac{1}{3}\operatorname{cosec} x+c\operatorname{cosec}^4 x.$$ □

***Example* 3**: Solve the equation $x\frac{dy}{dx}+2(y-4x^2)=0$.

Solution: Putting the equation in the standard form, we get $\frac{dy}{dx}+\frac{2}{x}y=8x$

The integrating factor is $\mu=e^{\int\frac{2}{x}dx}=x^2$, and the solution is

$$y=\frac{1}{x^2}\left\{\int x^2\cdot 8x\,dx+c\right\}=\frac{1}{x^2}(2x^4+c)$$

Or $y=2x^2+c/x^2$. □

***Example* 4**: Solve the equation $y\,dx+(3x-xy+2)\,dy=0$.

Solution: Putting the equation in the standard form and noting that x and y are interchanged, we get

$$\frac{dx}{dy}+(\frac{3}{y}-1)x=-\frac{2}{y}$$

The integrating factor is $\mu=e^{\int(\frac{3}{y}-1)dy}=e^{3\ln y-y}=y^3e^{-y}$, and the solution is

$$x=\frac{e^y}{y^3}\left\{\int y^3e^{-y}\cdot\left(-\frac{2}{y}\right)dy+c\right\}.$$

Integrating by parts twice the solution becomes

$$xy^3=2y^2+4y+4+ce^y.$$ □

Exercise 1.6

a. Solve the following differential equations:

1. $x\frac{dy}{dx} - 3y = x^5$ **Ans**: $2y = x^5 + cx^3$

2. $2(2xy + 4y - 3)\,dx + (x+2)^2 dy = 0$ **Ans**: $y = 2(x+2)^{-1} + c(x+2)^{-4}$

3. $u\,dx + (1-3u)x\,du = 3u^2 e^{3u}\,du$ **Ans**: $xu = (u^3 + c)e^{3u}$

4. $y' = \operatorname{cosec} x + y\cot x$ **Ans**: $y = c\sin x - \cos x$

5. $y' = x - 4xy$ **Ans**: $4y = 1 + ce^{-2x^2}$

6. $y' - my = e^{mx}$ **Ans**: $y = (x + c)e^{mx}$

7. $v\,dx + (2x + 1 - vx)\,dv = 0$ **Ans**: $xv^2 = v + 1 + ce^v$

8. $1/y' - x/y = 2y^2$ **Ans**: $x/y = y^2 + c$

9. $y' = 1 + 3y\tan x$ **Ans**: $3y\cos^3 x = c + \sin x - \sin^3 x$

10. $(y - x + xy\cot x)\,dx + x\,dy = 0$ **Ans**: $xy\sin x = c + \sin x - x\cos x$

11. $\sin\theta\frac{dr}{d\theta} = -1 - 2r\cos\theta$ **Ans**: $r\sin^2\theta = c + \cos\theta$

12. $\sin x\frac{dy}{dx} + 3y = \cos x$ **Ans**: $(y + \frac{1}{3})\tan^3(x/2) = 2\tan(x/2) - x + c$

13. $\sin x\frac{dy}{dx} + y\cos x = 2\sin^2 x\cos x$ **Ans**: $3y\sin x = 2\sin^3 x + c$

14. $(x + 2y^3)\frac{dy}{dx} = y$ **Ans**: $x = \tan^{-1} y - 1 + ce^{-\tan^{-1} y}$

15. $\frac{dy}{dx} - y\cot x = e^x(1 - \cot x)$ **Ans**: $y = e^x + c\sin x$

16. $\frac{dy}{dx} + y = \cos x - \sin x$ **Ans**: $y = \cos x + ce^{-x}$

17. $\frac{dy}{dx} + \frac{y}{x} = \cos x + \frac{\sin x}{x}$ **Ans**: $y = \sin x + \frac{c}{x}$

18. $x\frac{dy}{dx} - y = (x - 1)e^x$ **Ans**: $y = e^x + cx$

19. $x\dfrac{dy}{dx} - y = \cos\left(\dfrac{1}{x}\right)$ **Ans:** $y = cx - x\sin\left(\dfrac{1}{x}\right)$

20. $\dfrac{dy}{dx} + \dfrac{y}{x} = \sin x^2$ **Ans:** $yx = c - \dfrac{1}{2}\cos x^2$

21. $(x^2 - 1)\dfrac{dy}{dx} + 2xy = 1$ **Ans:** $y(x^2 - 1) = x + c$

b. Find the solutions that satisfy the given conditions:

1. $\dfrac{dy}{dx} - \dfrac{2y}{x} = x^2 e^x$; when $x = 1, y = 0$ **Ans:** $y = x^2(e^x - e)$

2. $\dfrac{dy}{dx} + \dfrac{2y}{x} = \dfrac{1}{x^2}$; when $x = 1, y = 2$ **Ans:** $y = \dfrac{x+1}{x^2}$

3. $\dfrac{dy}{dx} + y\tan x = \sec x$; when $x = 0, y = -1$ **Ans:** $y = \sin x - \cos x$

4. $(1+x^2)\dfrac{dy}{dx} + 2xy - 4x^2 = 0$, when $x = 0, y = 0$ **Ans:** $4x^3 = 3y(1+x^2)$

5. $y' + y\cot x = 5e^{\cos x}$; when $x = \dfrac{\pi}{2}, y = -4$ **Ans:** $y\sin x + 5e^{\cos x} = 1$

6. $(1+x)y' + (1+2x)y = (1+x)^2$; when $x = 0, y = 4$

   Ans: $y = \dfrac{1}{2}(1+x)(1+7e^{-2x})$

7. $\dfrac{dy}{dx} + \dfrac{2x}{1+x^2}y = \dfrac{1}{(1+x^2)^2}$; when $x = 1, y = 0$

   Ans: $y(1+x^2) = \tan^{-1}x - \dfrac{\pi}{4}$

8. $\dfrac{dy}{dx} - \dfrac{2y}{1+x} = (1+x)^3$, when $x = 0, y = 1$ **Ans:** $2y = (1+x)^4 + (1+x)^2$

c. Find the equation of the curve whose slope at any point is equal to $\dfrac{2y + x + 1}{x}$ and passes through the point $(1, 0)$. **Ans:** $2y = 3x^2 - 2x - 1$

d. Find the equation of the curve whose slope at any point is equal to $\dfrac{y^2\ln x - y}{x}$ and passes through the point $(1, 1)$. **Ans:** $y(\ln x + 1) = 1$

e. Find the solution of $y' = 2(2x - y)$ which passes through the point (0,-1).

   Ans: $y = 2x - 1$

1.3.6. Bernoulli's Equation

A well-known equation, which can fit into the previous category (linear equations), with some modification, is the Bernoulli's equation. This equation takes the form

$$\frac{dy}{dx}+P(x)y=Q(x)y^n \tag{23}$$

If $n = 1$, then the equation is solved by separation of variables; whereas, if $n = 0$, it reduces to a linear equation. Otherwise, we can put the equation in the form

$$y^{-n}\frac{dy}{dx}+Py^{-n+1}=Q,$$

or

$$\frac{1}{1-n}\cdot\frac{d}{dx}(y^{1-n})+Py^{1-n}=Q. \tag{24}$$

Now, let $y^{1-n}=z$, we obtain

$$\frac{dz}{dx}+(1-n)Pz=(1-n)Q, \tag{25}$$

which is a linear equation in z.

***Example* 1**: Solve the equation $\dfrac{dy}{dx}-y=xy^5$.

Solution: Dividing by y^5, we get

$$y^{-5}\frac{dy}{dx}-y^{-4}=x \quad \text{or} \quad -\frac{1}{4}\frac{d}{dx}(y^{-4})-y^{-4}=x.$$

Letting $y^{-4}=z$, the $\dfrac{dz}{dx}+4z=-4x$.

This is a linear equation in z. The integrating factor is

$\mu=e^{\int 4dx}=e^{4x}$, and

$$z=e^{-4x}\left\{\int e^{4x}\cdot x\,dx+x\right\}=-x+\frac{1}{4}+ce^{-4x}.$$

The solution is $\dfrac{1}{y^4}=-x+\dfrac{1}{4}+ce^{-4x}$. □

***Example* 2**: Solve the differential equation $\dfrac{dy}{dx}-y\tan x=\dfrac{\sin x}{y^2}$.

Solution: Multiplying by y^2, we get $y^2\dfrac{dy}{dx}-y^3\tan x=\sin x$, or

$$\frac{d}{dx}(y^3)-(3\tan x)y^3 = 3\sin x \ .$$

Letting $y^3 = z$, then $\dfrac{dz}{dx}-(3\tan x)z = 3\sin x$.

The integrating factor is $\mu = e^{-\int 3\tan x\, dx} = e^{-3\ln\sec x} = \cos^3 x$,

and $z\cos^3 x = \int 3\sin x\cos^3 x\, dx + c = -\dfrac{3}{4}\cos^4 x + c$.

The solution is $y^3 = -\dfrac{3}{4}\cos x + c\sec^3 x$. □

***Example* 3**: Find the solution of $\dfrac{dy}{dx}-\dfrac{1}{x}y = (1+\ln x)y^3$.

Solution: Dividing by y^3 , we get $y^{-3}\dfrac{dy}{dx}-\dfrac{1}{x}y^{-2} = (1+\ln x)$, or

$$-\frac{1}{2}\frac{d}{dx}(y^{-2})-\frac{1}{x}y^{-2} = (1+\ln x).$$

Letting $y^{-2} = z$, then $\dfrac{dz}{dx}+\dfrac{2}{x}z = (1+\ln x)$.

The integrating factor is $\mu = e^{\int \frac{2}{x}dx} = x^2$, and

$$z\,x^2 = \int -x^2(1+\ln x)dx = -\frac{2}{3}x^3\left(\frac{2}{3}+\ln x\right)+c .$$

The solution is $\dfrac{x^2}{y^2} = -\dfrac{2}{3}x^3\left(\dfrac{2}{3}+\ln x\right)+c$. □

Note: Bernoulli's equation can also be solved by assuming a solution of the form $y = uv$, where u and v are functions of x alone. The following example illustrates the procedure.

***Example* 4**: Solve the differential equation $\dfrac{dy}{dx}+\dfrac{1}{x}\cdot y = \ln x\cdot y^2$.

Solution: This is clearly Bernoulli's equation with $P(x) = \dfrac{1}{x}$, $Q(x) = \ln x$ and $n = 2$. Assume that the solution takes the form $y = uv$. If this is true it must satisfy the differential equation. Then

$$u\frac{dv}{dx}+v\frac{du}{dx}+\frac{1}{x}\cdot uv = \ln x\cdot u^2v^2 \text{, or}$$

$$u\frac{dv}{dx}+\left(\frac{du}{dx}+\frac{1}{x}\cdot u\right)v=\ln x\cdot u^2v^2 \tag{26}$$

We now choose u such that $\frac{du}{dx}+\frac{1}{x}\cdot u=0$.

Integrating, we obtain $u=\frac{1}{x}$. Equation (26) reduces to

$\frac{1}{x}\cdot\frac{dv}{dx}=\ln x\cdot\frac{1}{x^2}v^2$, or $\frac{dv}{v^2}=\frac{\ln x}{x}dx$.

Integrating, we get $-\frac{1}{v}=\frac{\ln^2 x}{2}+c$, or $v=-\frac{2}{\ln^2 x+2c}$.

The general solution of the differential equation is

$y=-\frac{1}{x}\cdot\frac{2}{\ln^2 x+2c}$, or $xy(\ln^2 x+2c)+2=0$. □

Note: A special equation that reduces to a linear equation is of the form $f'(y)\frac{dy}{dx}+Pf(y)=Q$, where P and Q are either constants or functions of x alone. To solve this type of equation, we let $v=f(y)$ so that $f'(y)\frac{dy}{dx}=\frac{dv}{dx}$, and the equation become $\frac{dv}{dx}+Pv=Q$. This is clearly a linear equation in v. Its integrating factor is $\mu=e^{\int P\,dx}$, and its solution is $v=\frac{1}{\mu}\left\{\int\mu Q\,dx+c\right\}$. The solution of the original equation is now $f(y)=\frac{1}{\mu}\left\{\int\mu Q\,dx+c\right\}$.

***Example* 5**: Solve the equation $\sec^2 y\frac{dy}{dx}+2x\tan y=x^3$.

Solution: Let $v=\tan y$, then $\sec^2 y\frac{dy}{dx}=\frac{dv}{dx}$. The equation reduces to

$\frac{dv}{dx}+2xv=x^3$. The integrating factor is $\mu=e^{\int 2x\,dx}=e^{x^2}$.

The solution is $e^{x^2}v=\int e^{x^2}x^3dx+c=\frac{1}{2}e^{x^2}(x^2-1)+c$.

Check the integration if you want!. The solution of the original equation is $\tan y=\frac{1}{2}(x^2-1)+ce^{-x^2}$. □

Exercise 1.7

a. Find the general solution of the following differential equations:

1. $y'+2xy+xy^4=0$ **Ans:** $\dfrac{1}{y^3}=-\dfrac{1}{2}+ce^{3x^2}$

2. $y'-y\tan x=-y^2\sec x$ **Ans:** $\sec x=y(c-\tan x)$

3. $y'+y=y^2(\cos x-\sin x)$ **Ans:** $1/y=-\sin x+ce^{-x}$

4. $xy'+y=y^2\ln x$ **Ans:** $1/y=\ln x+1+cx$

5. $(3\sin y-5x)dx+2x^2\cot y\,dy=0$ **Ans:** $x^3(\sin y-x)^2=c\sin^2 y$

6. $2xy\,dy-(x^2+y^2+1)\,dx=0$ **Ans:** $y^2=x(c+x)-1$

7. $(1-x^2)y'+xy=xy^2$ **Ans:** $cy=(1-y)\sqrt{1-x^2}$

8. $\dfrac{dv}{du}=(u-v)^2-2(u-v)-2$ **Ans:** $(u-v-3)e^{4u}=c(u-v+1)$

9. $yy'+y^2\tan x=\cos^2 x$ **Ans:** $y^2=(2x+c)\cos^2 x$

10. $(1+e^{y/x})dy+(1-y/x)dx=0$ **Ans:** $y+x\,e^{y/x}=c$

11. $xy'+y=y^2\ln x$ **Ans:** $y(\ln x+1+cx)=1$

12. $y'-y\tan x+y^2\sec x=0$ **Ans:** $\sec x=y(c-\tan x)$

13. $2y'=y/x+(y^2/x^2)$ **Ans:** $x=y(1+c\sqrt{x})$

14. $nx\dfrac{dy}{dx}+2y=xy^{n+1}$ **Ans:** $cx^2y^n+xy^n=1$

15. $x\dfrac{dy}{dx}+y=xy^3$ **Ans:** $cx^2y^2+2xy^2=1$

16. $\dfrac{dy}{dx}+\dfrac{xy}{1-x^2}-x\sqrt{y}=0$ **Ans:** $\sqrt{y}=-\tfrac{1}{3}(1-x^2)-c(1-x^2)^{1/4}$

17. $\dfrac{dy}{dx}+x\sin 2y=x^3\cos^2 y$ **Ans:** $\tan y=\tfrac{1}{2}(x^2-1)+ce^{-x^2}$

18. $\dfrac{dy}{dx}=e^{x-y}(e^x-e^y)$ **Ans:** $e^y=e^x-1+ce^{-e^x}$

19. $\dfrac{dy}{dx}+\dfrac{y}{x}\ln y=\dfrac{y}{x^2}\ln^2 y$ **Ans:** $\dfrac{1}{x\ln y}=\dfrac{1}{2x^2}+c$

20. $\sin y\,\dfrac{dy}{dx}=\cos y\,(1-x\cos y)$ **Ans:** $\sec y=x+1+c\,e^x$

21. $\dfrac{dy}{dx}-\dfrac{\tan y}{1+x}=(1+x)e^x\sec y$ **Ans:** $\sin y=(1+x)(c+e^x)$

22. $x\dfrac{dy}{dx}-y+xy^2=0$ **Ans:** $2x=y(x^2+c)$

23. $\dfrac{dy}{dx}+\dfrac{y}{x}=\dfrac{y^2}{x^2}$ **Ans:** $2x=cx^2y+y$

24. $y(x^2y+e^x)dx=e^x\,dy$ **Ans:** $x^3+c=-\dfrac{3e^x}{y}$

25. $\dfrac{dy}{dx}+xy=x^3y^4$ **Ans:** $\dfrac{1}{y^3}=x^2+\dfrac{2}{3}+c\,e^{3x^2/2}$

26. $\dfrac{dy}{dx}+y\cos x=y^n\sin 2x$ **Ans:** $y^{1-n}=2\sin x-\dfrac{2}{1-n}+c\,e^{-(1-n)\sin x}$

27. $\dfrac{dy}{dx}+y\cot x=y^2\sin^2 x\cos^2 x$ **Ans:** $\dfrac{1}{y\sin x}=\dfrac{1}{3}\cos^3 x+c$

b. Find the particular solution of the following equations:

1. $x^3y'=y(3y-x^2)$; when $x=1, y=1$ **Ans:** $y=x^2$

2. $y'=2(3x+y)^2-1$; when $x=0, y=1$ **Ans:** $4\tan^{-1}(3x+y)=8x+\pi$

3. $(y^4-2xy)\,dx+3x^2dy=0$; when $x=2, y=1$ **Ans:** $x^2=y^3(x+2)$

4. $2xyy'=y^2-2x^3$, when $x=1, y=2$ **Ans:** $y^2=x(5-x^2)$

5. $\dfrac{dy}{dx}=\dfrac{x^2+y^2+1}{2xy}$, when $x=1, y=1$ **Ans:** $y^2=x(1+x)-1$

c. Using the substitution $v=f(y)$, show that the equation

$f'(y)\dfrac{dy}{dx}+f(y)P(x)=Q(x)$ reduces to a linear equation in v. Then solve

the equation $\dfrac{dy}{dx}+1=4e^{-y}\sin x$ **Ans:** $e^y=2(\sin x-\cos x)+c\,e^{-x}$

1.3.7. Orthogonal Trajectories

Suppose that we have a family of curves defined by

$$\text{(I)} \quad f(x, y, c) = 0 \tag{27}$$

where c is a parameter. For each value of the parameter c corresponds a curve in this family (I). Now, we wish to determine another family (II) of curves, i.e.

$$\text{(II)} \quad g(x, y, k) = 0 \tag{28}$$

such that at any intersection of a curve of the family (II) with a curve of the family (I), the tangents to the two curves are perpendicular. The families (I) and (II) are then said to be ***orthogonal trajectories*** of each other.

To obtain the second family of curves (II) defined by equation (28), we know that, at the point of intersection, if the slope of the curve from family (I) is *dy/dx*, then the slope of the curve from family (II) has to be *-dx/dy*. From this reasoning, we have the following procedure:

1. Obtain the differential equation representing family (I) by differentiating equation (27) with respect to x.
2. Replace *dy/dx* by *-dx/dy* to obtain the differential equation representing family (II).
3. Solve the new differential equation to obtain the second family of curves (II) representing the orthogonal trajectories.

The following examples illustrate the procedure.

***Example* 1**: Find the orthogonal trajectories of all parabolas with vertices at the origin and foci on the *x-axis*.

Solution: The family (I) is $y^2 = 4ax$ or $\dfrac{y^2}{x} = 4a$. Differentiating with respect to x, we get $2x\dfrac{dy}{dx} - y = 0$.

Replacing *dy/dx* by *-dx/dy*, we get $2x\,dx + y\,dy = 0$.

Solving this differential equation, we obtain $2x^2 + y^2 = b^2$, where b is an arbitrary constant. Thus the orthogonal trajectories of the given family of parabolas are a family of ellipses with center at the origin. □

***Example* 2**: Find the orthogonal trajectories for $x^2 + (y - \beta)^2 = \beta^2$.

Solution: Differentiating with respect to x, we get $2x + 2(y - \beta)y' = 0$.

Eliminating β from the previous equations, we get $y' = \dfrac{2xy}{x^2 - y^2}$.

Replacing y' by $-1/y'$, we get $y' = -\dfrac{x^2 - y^2}{2xy}$.

This is a homogeneous equation, the solution of which is

$(x-\alpha)^2 + y^2 = \alpha^2$.

Then the orthogonal trajectories of the circles with radii β and centers at $(0, \beta)$ are circles with radii α and centers at $(\alpha, 0)$. □

Note: To obtain trajectories that intersect a given family at a constant angle α, we replace $p = \dfrac{dy}{dx}$ by $\dfrac{p + \tan\alpha}{1 - p\tan\alpha}$. Can you prove that?!.

Exercise 1.8

a. Find the orthogonal trajectories of the following curves:

1. $x^2 = 4\lambda(y+\lambda)$ **Ans:** $x^2 = 4\alpha(y+\alpha)$
2. $x^2 + y^2 + cx = 0$ **Ans:** $x^2 + y^2 = by$
3. $x = y + ce^y$ **Ans:** $y = x + 2 + be^x$
4. $ay^2 = x^3$ **Ans:** $3y^2 + 2x^2 = 2b$
5. $x^2 - y^2 = ax$ **Ans:** $y(y^2 + 3x^2) = b$
6. $x^n + y^n = a^n,\ n \neq 2$ **Ans:** $x^{2-n} - y^{2-n} = b$
7. $y = cx^3$ **Ans:** $x^2 + 3y^2 = b$
8. $e^x \sin y = c$ **Ans:** $e^x \cos y = b$
9. $y = ce^{x^2}$ **Ans:** $x = be^{-y^2}$

b. Find the differential equation of the family of curves given by the equation $x^2 - y^2 + 2\lambda x = 1$. Obtain the differential equation representing the orthogonal trajectories then solve it. **Ans:** $(x^2 + y^2)^2 - 2(x^2 - y^2) = c$

c. In polar coordinates, to obtain the orthogonal trajectories, we replace $r\dfrac{d\theta}{dr}$ in the differential equation by $-\dfrac{1}{r}\dfrac{dr}{d\theta}$. Find the orthogonal trajectories for the cardoids $r = a(1-\cos\theta)$. **Ans:** $r = c(1+\cos x)$

d. Find the family of curves whose tangents form a constant angle of $\pi/4$ with the family of hyperbolas $xy = c$. **Ans:** $y^2 - 2xy - x^2 = c$

e. Given the family of straight lines $y = ax$, find the family of curves that intersect these lines at a constant angle α.

Ans: $(x^2 + y^2)^{\tan x} = c\, e^{-2\tan^{-1}(y/x)}$

Chapter Two

Equations of First Order and Not of First Degree

$$xyp^2 + (x + y)p + 1 = 0$$

Chapter 2.

Equations of First Order and Not of First Degree

2.1. Introduction

In the previous chapter, we studied equations of first degree, i.e, the first derivative has power one. We now consider equations of the general form

$$f\,(x\,,y\,,y') = 0 \tag{1}$$

where $y\,'$ is of power higher than one. For the sake of simplification, we let

$$\frac{dy}{dx} = y\,' = p$$

2.2. Equations Solvable for p

In this case, it is possible to factor the equation into linear factors of the form

$$[p - f_1(x\,,y\,)][p - f_2(x\,,y\,)]\cdots[p - f_n(x\,,y\,)] - 0\,,$$

and then letting each linear factor equal to zero. Each linear factor when equated to zero will represent a first order differential equation which are of the first degree and can be solved using previous techniques. The general solution will be the product of all partial solutions. The following examples illustrate the procedure for the solution.

***Example* 1**: Solve the equation $xyp^2 + (x + y\,)p + 1 = 0$.

Solution: Factoring the left hand side of the equation we get

$(xp + 1)\,(yp + 1) = 0$.

Then, we have $xp + 1 = 0$ or $yp + 1 = 0$.

Hence $dy = -\dfrac{dx}{x}$ or $y\;dy = -dx$

Integrating, we get $y = -\ln\,(c\,x\,)$ or $y^2 = -2\,(x - c)$.

Then, the general solution can be written as

$[y + \ln\,(cx\,)][y^2 + 2(x - c)] = 0$ □

***Example* 2**: Solve the differential equation $p^2 - 2p\cosh x + 1 = 0$.

Solution: Solving the equation for *p*, we get $p = \cosh x \pm \sinh x = e^{\pm x}$, then

$\dfrac{dy}{dx} = e^x$ or $\dfrac{dy}{dx} = e^{-x}$.

Hence $y = e^x + c$ or $y = -e^{-x} + c$.

Then the general solution is

$$(y - e^x - c)(y + e^{-x} - c) = 0.$$ □

***Example* 3**: A curve whose differential equation is $p^2 + 2py\cot x - y^2 = 0$ passes through the point $(\pi/2, 1)$. Find the equation of this curve.

Solution: The equation of the curve is the solution of the differential equation with the condition that $y = 1$ when $x = \pi/2$. Solving for *p*, we get

$$p = \frac{-2y\cot x \pm \sqrt{4y^2\cot^2 x - 4y^2}}{2} = -y(\cot x \pm \operatorname{cosec} x).$$

Now we have two equations:

$$\frac{dy}{dx} = -y(\cot x + \operatorname{cosec} x) \quad \text{and} \quad \frac{dy}{dx} = -y(\cot x - \operatorname{cosec} x).$$

These two equations can be solved by separation of variables:

$$\frac{dy}{y} = -(\cot x + \operatorname{cosec} x)\,dx \quad \text{and} \quad \frac{dy}{y} = -(\cot x - \operatorname{cosec} x)\,dx$$

Integrating, we get $\ln y = -\ln\sin x - \ln(\operatorname{cosec} x - \cot x) + \ln c_1$, or

$$y = c_1\frac{\operatorname{cosec} x}{\operatorname{cosec} x - \cot x} = c_1\frac{1}{1 - \cos x} = 2c_1\operatorname{cosec}^2\frac{x}{2}.$$

Similarly we have $\ln y = -\ln\sin x + \ln(\operatorname{cosec} x - \cot x) + \ln c_2$, or

$$y = c_2\frac{\operatorname{cosec} x - \cot x}{\sin x} = 2c_2\sec^2\frac{x}{2}.$$

Then the general solution is

$$\left(y - c\operatorname{cosec}^2\frac{x}{2}\right)\left(y - c\sec^2\frac{x}{2}\right) = 0.$$

Note that there is no loss of generality if we take $2c_1 = 2c_2 = c$, since they are all arbitrary constants.

The equation of the curve that passes through $(\pi/2, 1)$ is found by $(1 - 2c)(1 - 2c) = 0$ or $c = 1/2$, then

$$\left(2y - \operatorname{cosec}^2\frac{x}{2}\right)\left(2y - \sec^2\frac{x}{2}\right) = 0$$ □

2.3. Equations Solvable for y

These equations reduce to the form

$$y = f(x, p) \tag{2}$$

From this we can interpret the equation as follows: y is a function of x and the "parameter" p. If we can obtain another equation of the form

$$y = \phi(p), \tag{3}$$

then we can say that the solution is obtained in parametric form given by Equations (2) and (3). Moreover, if we can eliminate the parameter p from the two equations, we obtain the cartesian form for the general solution.

***Example* 1**: Solve the equation $y = p + p^4$.

Solution: Differentiating with respect to x, we get $p = (1+4p^3)\dfrac{dp}{dx}$.

Separating the variables and integrating, we obtain

$$\int dx = \int \frac{1+4p^3}{p}\,dp = \ln p + \frac{4}{3}p^3 + c .$$

The general solution is $x = \ln p + \frac{4}{3}p^3 + c$ and $y = p + p^4$,

where p is now a parameter. □

***Example* 2**: Find the solution of the differential equation $xy'^2 - 2yy' + x = 0$.

Solution: Solving for y, we get $y = \dfrac{xp}{2} + \dfrac{x}{2p}$.

Differentiating with respect to x, we obtain

$$p = \left(\frac{x}{2} - \frac{x}{2p^2}\right)\frac{dp}{dx} + \frac{p}{2} + \frac{1}{2p} \quad \text{or} \quad (p^2-1)\left(x\frac{dp}{dx} - p\right) = 0 .$$

We obtain two equations

$$p^2 - 1 = 0 \quad \text{or} \quad x\frac{dp}{dx} - p = 0 .$$

Consider the second equation, we get by integration $p = cx$. Then the general solution is $y = \dfrac{xp}{2} + \dfrac{x}{2p}$ and $x = \dfrac{p}{c}$.

Eliminating p, we get the general solution in cartesian form as

$$y = \frac{cx^2}{2} + \frac{1}{2c}, \quad c \neq 0 .$$

□

Note: Equation $p^2 - 1 = 0$ gives the solutions $y = \pm x$. These are clearly solutions of the differential equations. They are called singular solutions. Singular solutions cannot be obtained from the general solution by any choice of the constant.

2.4. Equations Solvable for *x*

These equations reduce to the form

$$x = f(y, p) \tag{4}$$

The same reasoning mentioned in the previous subsection applies but interchanging *x* and *y*.

***Example* 1**: Solve the differential equation $x = yp + \frac{2y}{p}$.

Solution: Differentiating with respect to *y*, we get

$$\frac{1}{p} = p + \frac{2}{p} + (y - \frac{2y}{p^2})\frac{dp}{dy}.$$

Separating the variables, we obtain $\frac{dy}{y} = -\frac{p^2 - 2}{p(1+p^2)}dp$.

Using partial fraction and integrating, we get $y = cp^2(1+p^2)^{-3/2}$.

The general solution in parametric form (*p* the parameter) is given by

$x = yp + \frac{2y}{p}$ and $y = cp^2(1+p^2)^{-3/2}$. □

***Example* 2**: Solve the differential equation $y^2 \ln y = xpy + p^2$.

Solution: Solving for *x*, we get $x = \frac{y \ln y}{p} - \frac{p}{y}$.

Differentiating with respect to *y* and rearranging, we get

$$\left(\frac{y \ln y}{p^2} + \frac{1}{y}\right)\left(\frac{p}{y} - \frac{dp}{dy}\right) = 0.$$

Omitting the first factor, we obtain

$$\frac{p}{y} - \frac{dp}{dy} = 0 \text{ or } \frac{dp}{p} = \frac{dy}{y}.$$

Integrating, $\ln p = \ln y + \ln c$ or $p = cy$.

Finally, the solution becomes $x = \frac{\ln y}{c} - c$. □

2.5. Clairaut's Equation

An equation of the form

$$y = xp + f(p), \qquad (5)$$

where $f(p)$ does not contain x or y explicitly, is called **Clairaut's equation**. Clearly this equation can be solved using the technique of section 4.2. (Equations solvable for y). But, because the solution of this equation is somewhat special, we treat it separately here.

Differentiating with respect to x, we get

$$p = p + x\frac{dp}{dx} + f'(p)\frac{dp}{dx}, \text{ or } [x + f'(p)]\frac{dp}{dx} = 0, \text{ then}$$

$$\frac{dp}{dx} = 0 \text{ or } p = c.$$

And the solution is $\qquad y = cx + f(c)$

We can see that the solution of Clairaut's equation is simply obtained by replacing p by an arbitrary constant c. This solution represents the equation of a family of straight lines. On the other hand, if we consider the equation

$$x + f'(p) = 0$$

together with he original Clairaut's equation, we have

$$y = f(p) - pf'(p).$$

The last two equations also represent the parametric solution for the Clairaut's equation. And if $f(p)$ is not linear in p or a constant, then the parametric solution cannot be obtained from the general solution by any choice of the constant c. Hence this parametric solution is in fact a ***singular solution*** for Clairaut's equation.

Example **1**: Solve the equation $y = px + 2p^2$.

Solution: Clearly this is a Clairaut equation with $f(p) = 2p^2$. then the general solution is $y = cx + 2c^2$.

To obtain the singular solution, we have $f'(p) = 4p$, and

$x = -4p$ and $y = px + 2p^2 = -2p^2$.

Eliminating p from the last two equations, we get $\quad y = -x^2/8$ □

Example **2**: Solve the equation $y = px + \frac{1}{p}$, and find the singular solution.

Solution: This is a Clairaut's equation with $f(p) = 1/p$. then the general solution is $y = cx + 1/c$

To obtain the singular solution, we have $f'(p) = -\dfrac{1}{p^2}$, and

$x = \dfrac{1}{p^2}$ and $y = px + \dfrac{1}{p}$.

Eliminating p from the last two equations, we get

$y = 2\sqrt{x}$ or $y^2 = 4x$. □

Note: The last two examples reveal that in the case of Clairaut's equation, the singular solution represents the envelope of the integral curves given by the general solution of the differential equation.

Exercise 2.1

Solve the following equations:

1. $x^2p^2 - y^2 = 0$ **Ans:** $(y - cx)(xy - c) = 0$

2. $xp^2 + (1 - x^2y)p - xy = 0$ **Ans:** $(y - c\,e^{x^2/2})(y + \ln(cx)) = 0$

3. $yp^2 - (x - y^2)p - xy = 0$ **Ans:** $(x^2 + y^2 - c)(y - c\,e^x) = 0$

4. $xy(x^2 + y^2)(p^2 - 1) = p(x^4 + x^2y^2 + y^4)$

   Ans: $(y^2 + 2xy - c)(y^2 + 2xy - x^2 - c) = 0$

5. $xp^3 - (x^2 + x + p)p^2 + (x^2 + xy + y)p - xy = 0$

   Ans: $(y - c\,x)(y - x - c)(x^2 - 2(y - c)) = 0$

6. $p^2 + p(x + y) + xy = 0$ **Ans:** $(2y + x^2 - c)(x + \ln y - c) = 0$

7. $p^2 + p(x^2y - xy) - x^3y^2 = 0$ **Ans:** $(y - c\,e^{x^2/2})(y + c\,e^{-x^3/3}) = 0$

8. $2p^2 + x^3p - 2x^2y = 0$ **Ans:** $y = c\,x^2 + 4c^2$; sing. sol.: $y = -x^4/6$

9. $xp^2 - yp - xy = 0$ **Ans:** $x = c(p - x)^3 - p$

10. $y - 2px - p^2y = 0$ **Ans:** $2cx - y^2 + c^2 = 0$

11. $y^2 \ln y = xpy + p^2$ **Ans:** $\ln y = cx + c^2$

12. $p^3 - 4xyp + 8y^2 = 0$ **Ans:** $y = c(c - x)^2$

13. $y = -px + x^4 p^2$ **Ans:** $y = c^2 - c/x$

14. $y = 2px + p^4 x^2$ **Ans:** $(y - c^2)^2 = 4cx$

15. $yp^2 + (2x - 1)p - y = 0$ **Ans:** $c^2 + c(2x - 1) + y^2 = 0$

16. $y = 2px + y^2 p^3$ **Ans:** $y^2 = 2cx + c^3$

17. $y = px + 2\sqrt{1 + p^2}$ **Ans:** $y = cx + 2\sqrt{1 + c^2}$

18. $p^2 + 3xp - y = 0$ **Ans:** $p^3(5x + 2p)^2 = c$ and $y = 3xp + p^2$

19. $x - 2p - \ln p = 0$ **Ans:** $x = 2p + \ln p,\ y = p^2 + p + c$

20. $xp^2 + yp = 3y^4$ **Ans:** $3y = c(1 + cxy)$; sing. sol.: $12xy^2 = -1$

21. $xp^2 - yp - y = 0$ **Ans:** $x = c(p + 1)e^p$ and $y = cp^2 e^p$

22. $y^3 p^2 - xp + y = 0$ **Ans:** $5y^2 = cp^{-4/3} - 2p^{-3}$, $5px = y(3 + cp^{3/2})$

23. $y^4 p^3 - 6xp + 2y = 0$ **Ans:** $y^3 = 18c(x - 6c^2)$; sing. sol.: $y^2 = 2x$

24. $p^2 + 2xy^3 p + y^4 = 0$ **Ans:** $4cy^2(x - c) = 1$; sing. sol.: $xy = -1$

25. $y = px + \ln p$ **Ans:** $y = cx + \ln c$; sing. sol. $xe^{y+1} + 1 = 0$

26. $p^2 = \sin^2 x$ **Ans:** $y = c \pm \cos x$

27. $e^{3x}(p - 1) + p^3 e^{2y} = 0$ **Ans:** $e^y = ce^x + c^3$

28. $y = px + \sqrt{1 + p^2}$ **Ans:** $y = cx + \sqrt{1 + c^2}$

29. $y = px + 2p(1 - p)$ **Ans:** $y = cx + 2c(1 - c)$

30. $y = px + p^n$ **Ans:** $y = cx + c^n$

Chapter Three

Second Order Differential Equations

$$a y'' + b y' + c y = 0$$

Chapter 3.

Second Order Differential Equations

3.1. Introduction

In general, higher order differential equations are more complicated and difficult to solve than first order. However, different approaches to solve some of these equations are available. The form of the differential equation dictates the method to be used. We confine our study with second order differential equations; however, the techniques presented in this chapter can sometimes be extended to cover higher order equations. We start with simple forms and then proceed to more involved ones.

3.2. Equations Reducible to First Order

3.2.1. Equations of the Form $\frac{d^2y}{dx^2} = f(x)$

These equations can be solved directly by integrating with respect to x twice.

$$\frac{dy}{dx} = \int f(x)dx + a$$

$$y = \int \left\{ \int f(x)dx + a \right\} dx + b$$

We notice here that in the solution y, two arbitrary constants a, b are present. If we are given two conditions, we can then find the corresponding solution by evaluating the arbitrary constants a and b that satisfy the given conditions (initial or boundary conditions).

***Example* 1**: Find the solution of the differential equation

$$\frac{d^2y}{dx^2} = 12x, \text{ when } x = 0, y = 0, \frac{dy}{dx} = 0.$$

Solution: Integrating once, we get $\frac{dy}{dx} = 6x^2 + a$.

Integrating again, we obtain the general solution as

$y = 2x^3 + ax + b$.

To satisfy the given conditions, we can see that $a = 0$ and $b = 0$. Then the solution is $y = 2x^3$. □

Example 2: Solve the equation $\dfrac{d^3y}{dx^3} = x\,e^x$.

Solution: Integrating once, we get

$$\frac{d^2y}{dx^2} = \int xe^x\,dx + c_1 = xe^x - e^x + c_1$$

Integrating one more time, we get

$$\frac{dy}{dx} = \int (xe^x - e^x + c_1)dx + c_2 = xe^x - 2e^x + c_1x + c_2$$

Again integrating one more time, we finally get

$$y = \int (xe^x - 2e^x + c_1x + c_2)dx = xe^x - 3e^x + \frac{1}{2}c_1x^2 + c_2x + c_3 .$$

3.2.2. Equations of the Form $\dfrac{d^2y}{dx^2} = g(y)$

In this case, we let $p = \dfrac{dy}{dx}$, then, $\dfrac{d^2y}{dx^2} = \dfrac{dp}{dx} = \dfrac{dp}{dy}\cdot\dfrac{dy}{dx} = p\,\dfrac{dp}{dy}$.

Example: Solve the differential equation $\dfrac{d^2y}{dx^2} = y$.

Solution: Let $\dfrac{dy}{dx} = p, \dfrac{d^2y}{dx^2} = p\,\dfrac{dp}{dy}$, then

$$p\frac{dp}{dy} = y \quad or \quad \int p\,dp = \int y\,dy .$$

Integrating, we get $p^2 = y^2 + x$ or $\dfrac{dy}{dx} = \pm\sqrt{y^2 + a}$.

Separating the variables and integrating, we obtain the general solution as $\ln\left(y + \sqrt{y^2 + a}\right) = \pm x + b$.

Note again the presence of the two arbitrary constants. □

3.2.3. Equations Not Containing the Dependent Variable

These equations take the form $\Psi\left(\dfrac{d^2y}{dx^2}, \dfrac{dy}{dx}, x\right) = 0$. Letting $\dfrac{dy}{dx} = p$, the equation reduces to $\Psi\left(\dfrac{dp}{dx}, p, x\right) = 0$. This is a first order differential equation in p, where x is the independent variable and p is the dependent variable.

Example: Solve the equation $x^2\dfrac{d^2y}{dx^2}+2x\dfrac{dy}{dx}-1=0$.

Solution: Let $\dfrac{dy}{dx}=p$, then $\dfrac{dp}{dx}+\dfrac{2}{x}\cdot p=\dfrac{1}{x^2}$.

This is a first order linear differential equation in p. The integrating factor is $\mu=e^{\int\frac{2}{x}dx}=e^{2\ln x}=x^2$, and

$$p=\frac{1}{x^2}\left\{\int\frac{x^2}{x^2}dx+a\right\}=\frac{1}{x}+\frac{a}{x^2}.$$

Integrating again, we obtain $y=\ln x-\dfrac{1}{x}+b$. □

3.2.4. Equations Not Containing the Independent Variable

These equations take the form $\Psi\left(\dfrac{d^2y}{dx^2},\dfrac{dy}{dx},y\right)=0$. Letting $\dfrac{dy}{dx}=p$ and $\dfrac{d^2y}{dx^2}=p\dfrac{dp}{dx}$, the equation reduces to $\Psi\left(\dfrac{dp}{dx},p,x\right)=0$. This is a first order differential equation, where y is now the independent variable and p the dependent variable.

Example: Solve the equation $y\dfrac{d^2y}{dx^2}=\left(\dfrac{dy}{dx}\right)^2+\dfrac{dy}{dx}$.

Solution: Let, $\dfrac{dy}{dx}=p$ and $\dfrac{d^2y}{dx^2}=p\dfrac{dp}{dx}$, then

$$yp\frac{dp}{dy}=p^2+p \quad \text{or} \quad \frac{dp}{p+1}=\frac{dy}{y}.$$

Integrating, we get $\ln(p+1)=\ln y+\ln a$ *or* $p+1=ay$.

Integrating again, the general solution is $x=a\ln(ay-1)+b$. □

Note: If the equation does not contain either the independent or the dependent variables, one of the previous two procedure can be used.

Example: Solve the equation $\dfrac{d^2y}{dx^2}=\dfrac{dy}{dx}$.

Solution: Let $\dfrac{dy}{dx}=p$, then $\dfrac{dp}{dx}=p$ or $\dfrac{dp}{p}=dx$.

Integrating, we get $x=\ln ap$ or $p=e^x/a$.

Integrating again, the solution is $y=e^x/a+b$. □

Exercise 3.1

a. Find the general solution of the following equations:

1. $xy''+y'+2x=0$ **Ans:** $y=-\frac{1}{2}x^2+a\ln x+b$

2. $(x^2-1)y''+xy'=0$ **Ans:** $y=a\ln\left(x+\sqrt{x^2-1}\right)+b$

3. $y''=\sec^2 x$ **Ans:** $y=\ln\sec x+ax+b$

4. $\frac{d^2y}{dx^2}=x^2\sin x$ **Ans:** $y=ax+b-x^2\sin x-4x\cos x+6\sin x$

5. $xy''=1+y'$ **Ans:** $y=ax^2-x+b$

6. $y^3y''=1$ **Ans:** $(c_1x+c_2)^2=c_1y^2-1$

7. $y''+y'=e^{-x}$ **Ans:** $y=ae^{-x}-xe^{-x}+b$

8. $xy'''=2y''$ **Ans:** $y=ax^4+bx+c$

9. $y^2\frac{d^2y}{dx^2}+4=0$ **Ans:** $4+ay^2=b^2(x+b)^2$

10. $\frac{d^4y}{dx^4}-2\frac{d^3y}{dx^3}=x^3$ **Ans:** $y=\frac{1}{100}x^6+ax^5+bx^2+cx+d$

11. $y'''=\ln x$ **Ans:** $36y=6x^3\ln x-11x^3+c_1x^2+c_2x+c_3$

12. $x^2\frac{d^4y}{dx^4}+1=0$ **Ans:** $y=\frac{1}{2}x^2\ln x+c_1x^3+c_2x^2+c_3x+c_4$

13. $x^2\left(\frac{d^2y}{dx^2}\right)^2=1+\left(\frac{dy}{dx}\right)^2$ **Ans:** $y=\frac{1}{2}\left(\frac{ax^2}{2}-\frac{1}{a}\ln x\right)+b$

14. $y(y-1)\frac{d^2y}{dx^2}+\left(\frac{dy}{dx}\right)^2=0$ **Ans:** $a(y-\ln y)=x+b$

15. $(1-x^2)y''-xy'=2$ **Ans:** $y=(\sin^{-1}x)^2+a\sin^{-1}x+b$

16. $y''+y'+y'^3=0$ **Ans:** $y=-\sin^{-1}(c_1e^{-x})+c_2$

17. $y'-xy''=f(y'')$ **Ans:** $y=\frac{1}{2}c_1x^2+xf(c_1)+c_2$

18. $\frac{d^4y}{dx^4}-\frac{d^2y}{dx^2}=0$ **Ans:** $y=c_1e^x+c_2e^{-x}+c_3x+c_4$

19. $\frac{d^4y}{dx^4}-\cot x\frac{d^3y}{dx^3}=0$ **Ans:** $y=c_1\cos x+c_2x^2+c_3x+c_4$

20. $yy''+1=y'^2$ **Ans:** $c_1y=\sinh(c_1x+c_2)$

21. $y^2y''=1$ **Ans:** $(c_1x+c_2)^2=c_1y^2-1$

22. $y(1-\ln y)y''+(1+\ln y)y'^2=0$ **Ans:** $(1-\ln y)(c_1x+c_2)=1$

23. $yy''+y'^2=1$ **Ans:** $y^2=x^2+c_1x+c_2$

b. Find the solution for the following initial-value problems:

1. $y''=\frac{1}{y^3}$, when $x=0, y=1, \frac{dy}{dx}=0$ **Ans:** $y^2-x^2=1$

2. $y''+y'=1$, when $x=0, y=0, \frac{dy}{dx}=0$ **Ans:** $y=x+e^{-x}-1$

3. $y''=e^x+6x$, when $x=0, y=0, \frac{dy}{dx}=1$ **Ans:** $y=e^x+x^3-1$

4. $y''=\sec^2 y\tan y$, when $x=0, y=0, \frac{dy}{dx}=1$ **Ans:** $y=\sin^{-1}x$

3.3. Linear Differential Equations

A linear second order differential equation takes the form

$$a_0(x)\frac{d^2y}{dx^2}+a_1(x)\frac{dy}{dx}+a_2(x)y=f(x) \qquad (1)$$

The coefficients of the function y and its first and second derivatives can be functions in the independent variable x. if $f(x)$ on the right hand side of Equation (1) is equal to zero, the equation is called ***homogeneous linear equation*** (not be to confused with the homogeneous first order differential equation). For Equation (1), if $f(x)$ is replaced by zero, the resultant equation is called the ***related homogeneous equation***.

If the coefficients $a_0(x)$, $a_1(x)$ and $a_2(x)$ are all constants (independent of x), then Equation (1) is said to have ***constant coefficients*** even through $f(x)$ on the right hand side may depend on x.

Examples of second order linear differential equations are

$$y''+4y'-5y=\sin 2x \quad \text{(constant coefficients)}$$

$$x^2y''-xy'+e^x y=\ln x \quad \text{(variable coefficients)}$$

$$x^2y''+xy'+y=0 \quad \text{(homogeneous equation)}$$

Note: Here, we will consider only the second order equation. However, the analysis can be easily extended to higher linear differential equations. The linear n^{th} order differential equation is of the form

$$a_0(x)\frac{d^ny}{dx^n}+a_1(x)\frac{d^{(n-1)}y}{dx^{(n-1)}}+\cdots+a_n(x)y=f(x)$$

Consider now the linear homogeneous second order differential equation

$$a_0(x)\frac{d^2y}{dx^2}+a_1(x)\frac{dy}{dx}+a_2(x)y=0 \qquad (2)$$

Assume that $y_1(x)$ and $y_2(x)$ are two solutions of Equation (2). Then it is clear that $y=c_1y_1+c_2y_2$ is also a solution (the proof is left to the reader). We do not know for now whether the two solutions are related or not. We need a definition.

Definition: Two functions $g_1(x)$ and $g_2(x)$ are said to be linearly dependent if there exists two constants b_1 and b_2 such that

$$b_1g_1(x)+b_2g_2(x)=0 \qquad (3)$$

If no such relation exists (i.e., b_1 and b_2 equal zero) then the two functions are said to be linearly independent.

To obtain a sufficient condition for linear independence over some interval, we proceed as follows. Differentiating Equation (3) with respect to x, we get

$$b_1 g_1'(x) + b_2 g_2'(x) = 0 \tag{4}$$

For the system of Equations (3) and (4) to have a non-trivial solution for b_1 and b_2, the determination of the system should not vanish, i.e.,

$$\begin{vmatrix} g_1 & g_2 \\ g_1' & g_2' \end{vmatrix} \neq 0 \tag{5}$$

In this case, the two functions are linearly independent. The mentioned determinant is called the ***Wronskian*** of the system of linear equations, i.e.,

$$W = \begin{vmatrix} g_1 & g_2 \\ g_1' & g_2' \end{vmatrix} \tag{6}$$

Two functions are linearly independent if their Wronskian W does not vanish. This can be extended to several functions as follows. Consider the equation

$$b_1 g_1(x) + b_2 g_2(x) + \cdots + b_n g_n(x) = 0 \tag{7}$$

It follows, by successive differentiation, that

$$b_1 g_1'(x) + b_2 g_2'(x) + \cdots + b_n g_n'(x) = 0$$

$$b_1 g_1''(x) + b_2 g_2''(x) + \cdots + b_n g_n''(x) = 0$$

$$b_1 g_1'''(x) + b_2 g_2'''(x) + \cdots + b_n g_n'''(x) = 0$$

$$\cdots\cdots\cdots\cdots\cdots\cdots\cdots\cdots$$

$$b_1 g_1^{(n-1)}(x) + b_2 g_2^{(n-1)}(x) + \cdots + b_n g_n^{(n-1)}(x) = 0$$

The Wronskian of this system is given by

$$W = \begin{vmatrix} g_1(x) & g_2(x) & \cdots & g_n(x) \\ g_1'(x) & g_2'(x) & \cdots & g_n'(x) \\ \cdots & \cdots & \cdots & \cdots \\ g_1^{(n-1)}(x) & g_2^{(n-1)}(x) & \cdots & g_n^{(n-1)}(x) \end{vmatrix} \tag{8}$$

Then, n functions are linearly independent if their Wronskian does not vanish.

Linear dependence or independence plays an important role in the solution of differential equations. This can be illustrated as follows. Consider the equation

$$y''' - y'' - y' + y = 0$$

We can verify that e^x , xe^x and e^{-x} are all solutions for this differential equation. Also the Wronskian of these three functions is e^{4x} , i.e. does not vanish for finite values of x. Thus, the three functions (solutions) are linearly independent.

Moreover, any linear combination of the form

$$y = c_1 e^x + c_2 x e^x + c_3 e^{-x}$$

is also a solution of the differential equation. In fact, since it contains three arbitrary constants, it is the general solution of the equation.

Exercise 3.2

a. Check the following functions for linear dependency:

1. e^{-x}, e^x, e^{2x}
2. $1, x, x^2, \quad x > 0$
3. $\ln x, \ln x^2, \ln x^3, \quad x > 0$
4. $\cosh^2 x, \sinh^2 x, 1, \quad x > 0$
5. $\sin x, \cos x, 1$
6. $e^x, e^{-x}, \cos x, \sin x$
7. $1+x, 1+2x, x^2$
8. $x^2 - x + 1, x^2 - 1, 3x^2 - x - 1$
9. $\sin 3x, \sin x, \sin^3 x$

b. Show that $y = x^2 + e^{-x}(c_1 \cos 2x + c_2 \sin 2x)$ is the general solution of the differential equation $y'' + 2y' + 5y = 5x^2 + 4x + 2$.

c. For the differential equation $x^2 y'' + 4xy' + 2y = 0$, verify that $y = c_1 x^{-2} + c_2 x^{-1}$ satisfies the equation for every choice of c_1 and c_2.

d. If $y_1(x)$ and $y_2(x)$ are solutions of the differential equation $y'' + p(x)y' + q(x)y = 0, \ a < x < b$, show that

1. $W'(x) = -p(x)W(x)$
2. $W(x) = ce^{-\int p(x)\,dx}$, where c is a constant.

3.4. Homogeneous Linear Equations with Constant Coefficients

We consider here the homogeneous linear second order differential equation whose coefficients are constants, namely

$$ay'' + by' + cy = 0 \tag{9}$$

We recall that the solution of the homogeneous linear first order differential equation

$$y' + ky = 0 \tag{10}$$

is an exponential function of the from $y = Ce^{-kx}$.

This suggests that we may try a similar solution for the second order equation. Assume that a solution of Equation (9) is of the form

$$y = e^{\lambda x}. \tag{11}$$

where λ is a constant to be determined. If this assumption is true, then this function must satisfy the differential equation (9). We have

$$y' = \lambda e^{\lambda x} \text{ and } y'' = \lambda^2 e^{\lambda x}$$

Substituting for y, y' and y'' in Equation (9), we get

$$e^{\lambda x}(a\lambda^2 + b\lambda + c) = 0 \tag{12}$$

and since $e^{\lambda x} \neq 0$ for finite values of x, then

$$\boxed{a\lambda^2 + b\lambda + c = 0} \tag{13}$$

This equation is a quadratic equation in λ, so it has two roots. Hence

$$\lambda_1 = \frac{-b + \sqrt{b^2 - 4ac}}{2a} \text{ and } \lambda_2 = \frac{-b - \sqrt{b^2 - 4ac}}{2a}.$$

Equation (13) is called the ***auxiliary equation*** or ***characteristic equation*** of the differential equation. The solution is now either $e^{\lambda_1 x}$ or $e^{\lambda_2 x}$ or any linear combination of these two solutions. We say that the general solution of the differential equation is

$$y = A_1 e^{\lambda_1 x} + A_2 e^{\lambda_2 x} \tag{14}$$

where A_1 and A_2 are any two arbitrary constants. They can be evaluated if we are given two conditions, *eg*. Initial conditions, boundary conditions.

So, the solution steps are summarized as follows.

1. Obtain the auxiliary equation as in Equation (13).
2. Solve the auxiliary equation to get the two roots λ_1 and λ_2
3. Write the general solution as in Equation (14).

Now, depending on the values of these roots, we may have:

1. λ_1 and λ_2 are real and distinct, the discriminant $b^2 - 4ac > 0$. the two solutions are linearly independent. The general solution is as given by Equation (14).
2. λ_1 and λ_2 are complex conjugate, i.e. $\lambda_{1,2} = \alpha \pm i\beta$. The discriminant $b^2 - 4ac < 0$. The two solutions are also linearly independent. The general solution will be in complex terms:

   $$y = A_1 e^{(\alpha + i\beta)x} + A_2 e^{(\alpha + i\beta)x} .$$

   In the Examples 2 and 3 below, we will see that the general solution can be put in real terms.
3. λ_1 and λ_2 are real and equal, the discriminant $b^2 - 4ac = 0$. Then, we have only one solution $y = A_1 e^{\lambda_1 x}$. To get the second solution, we proceed as follows. If y_1 in the first solution, we assume that the second solution takes the form $y_2 = uy_1$, where u is a function in x. If this is true, then this function y_2 must satisfy the differential equation. We have

   $y_2' = uy_2' + u'y_1$, and $y_2'' = uy_2'' + 2u'y_1' + u''y_1$.

   Substituting for y, y' and y'' in the differential equation and re-arranging, we get

   $$u(ay_1'' + by_1' + cy_1) + u'(2ay_1' + by_1) + u''y_1 = 0.$$

   But $ay_1'' + by_1' + cy_1 = 0$. Also, since we have a repeated root, then

   $$2ay_1' - by_1 = 0 .$$

   Hence we get $u''y_1 = 0$.

   Since $y_1 \neq 0$, then $u'' = 0$, i.e., $u = (Ax + B)$, and the second solution will be $y_2 = (Ax + B)y_1$. The general solution will be a linear combination of the two solution and is given by

   $$y = e^{\lambda x}(Cx + D)$$

   where λ is the repeated root and C and D are the two arbitrary constants.

***Example* 1**: Distinct real roots. Find the solution of $y'' + y' - 6y = 0$.

Solution: The auxiliary equation is $\lambda^2 + \lambda - 6 = 0$. Then the roots are $\lambda_1 = -3$ and $\lambda_2 = 2$. The general solution is

$$y = ae^{-3x} + be^{2x} .$$

***Example* 2**: Complex conjugate roots. Find the solution of $y''+y=0$.

Solution: The auxiliary equation is $\lambda^2+1=0$, and the roots are $\lambda_1=i$ and $\lambda_2=-i$. The solution is

$$y=ae^{ix}+be^{-ix}=a(\cos x+i\sin x)+b(\cos x+i\sin x)$$

$$=(a+b)\cos x+i(a-b)\sin x.$$

We can write the solution in real terms as $y=A\cos x+B\sin x$,

where A and B are now the two arbitrary constants.

***Example* 3**: Complex conjugate roots. Find the solution of $y''+2y'+2y=0$.

Solution: The auxiliary equation is $\lambda^2+2\lambda+2=0$, and the two roots are

$\lambda_1=-1+i$ and $\lambda_2=-1-i$.

The solution is
$y=ae^{(-1+i)x}+be^{(-1-i)x}=e^{-x}(ae^{ix}+be^{-ix})$.

Then, the general solution is $y=e^{-x}(A\cos x+B\sin x)$.

***Example* 4**: Real double roots. Solve the equation $y''-2y'+y=0$.

Solution: The auxiliary equation is $\lambda^2-\lambda+1=0$, and the roots are $\lambda_1=\lambda_2=1$. The general solution will be $y=(ax+b)e^x$.

***Example* 5**: Initial value problem. Solve the initial value problem:

$$y''-4y'+4y=0, \text{ when } x=0, y=3 \text{ and } y'=1.$$

Solution: The auxiliary equation is $\lambda^2-4\lambda+4=0$; then we have a real double root at $\lambda=2$. The general solution is $y=(ax+b)e^{2x}$.

To satisfy the give initial conditions the value of the to arbitrary constants a and b are $a=-5$ and $b=3$.

The solution of the initial value problem is $y=(3-5x)e^{2x}$.

To summarized, we give the following table.

***Table* 1.** Value of the discriminant.

Discriminant	Root of the auxiliary equation	General solution
$b^2-4ac>0$	Distinct real λ_1 and λ_2	$y=A\,e^{\lambda_1 x}+B\,e^{\lambda_2 x}$
$b^2-4ac<0$	Complex conjugate $\lambda_{1,2}=\alpha\pm i\beta$	$y=e^{\alpha x}\,(A\cos\beta x+B\sin\beta x)$
$b^2-4ac>0$	$\lambda_{1,2}=\alpha\pm\sqrt{\beta}$	$y=e^{\alpha x}\,(A\cosh\sqrt{\beta}x+B\sinh\sqrt{\beta}x)$
$b^2-4ac=0$	Double root λ	$y=(Ax+B)e^{\lambda x}$

Exercise 3.3

a. Find a differential equation such that y is its general solution:

1. $y=e^{-x}(A\cos x+B\sin x)$ **Ans:** $y''+2y'+2y=0$
2. $y=(c_1+c_2x)e^{4x}$ **Ans:** $y''-8y'+16y=0$
3. $y=A\cosh 3x+B\sinh 3x$ **Ans:** $y''-9y=0$
4. $y=c_1+c_2e^{4x}$ **Ans:** $y''-4y'=0$

b. Find the general solution of the following differential equation:

1. $y''+\pi^2y=0$ **Ans:** $y=A\cos\pi x+B\sin\pi x$
2. $y''+6y'+4y=0$ **Ans:** $y=e^{-3x}(a_1\cosh\sqrt{5}x+a_2\sinh\sqrt{5}x)$
3. $y''+2ay'+(a^2+b^2)y=0$ **Ans:** $y=e^{-ax}(a_1\cos bx+a_2\sin bx)$
4. $y''+y'+y=0$ **Ans:** $y=e^{-x/2}(a_1\cos\frac{\sqrt{3}}{2}x+a_2\sin\frac{\sqrt{3}}{2}x)$
5. $y''+6y'+9y=0$ **Ans:** $y=(a+bx)e^{-3x}$
6. $2y''+3y'-2y=0$ **Ans:** $y=ae^{x/2}+be^{-2x}$
7. $4y''-4y'+y=0$ **Ans:** $y=(a+bx)e^{x/2}$
8. $8y''-2y'-y=0$ **Ans:** $y=ae^{x/2}+be^{-x/4}$

9. $y''' + 6y'' + 12y' + 8y = 0$ **Ans:** $y = (a_1 + a_2 x + a_3 x^2)e^{-2x}$

10. $y''' - 2y'' + 4y' - 8y = 0$ **Ans:** $y = a_1 e^{2x} + a_2 \cos 2x + a_3 \sin 2x$

11. $y''' - 4y'' + 5y' - 2y = 0$ **Ans:** $y = (a_1 + a_2 x)e^{x} + a_3 e^{2x}$

12. $y''' + 6y'' + 11y' + 6y = 0$ **Ans:** $y = a_1 e^{-x} + a_2 e^{-2x} + a_3 e^{-3x}$

13. $y''' + 3y'' + 3y' + y = 0$ **Ans:** $y = (a_1 + a_2 x + a_3 x^2)e^{-x}$

14. $y'''' - 7y''' + 118y'' - 20y' + 8y = 0$

Ans: $y = (a_1 + a_2 x + a_3 x^2)e^{-x} + a_4 e^{4x}$

c. Solve the following initial-value problems:

1. $y'' - 4y = 0$, when $x = 0, y = 2$ and $y' = 4$ **Ans:** $y = 2e^{2x}$

2. $y'' - 6y' + 9y = 0$, when $x = 0, y = 2$ and $y' = 8$

Ans: $y = (2 + 2x)e^{3x}$

3. $y'' - 4y' + 13y = 0$, when $x = 0, y = -1$ and $y' = 2$

Ans: $y = e^{2x}(-\cos 3x + \frac{4}{3}\sin 3x)$

4. $y'' + 4y' + 5y = 0$, when $x = 0, y = 1$ and $y' = -3$

Ans: $y = e^{-2x}(\cos x - \sin x)$

5. $y'' + y = 0$, when $x = 0, y = 1$ and $y' = 0$ **Ans:**
$y = \cos x$

d. Solve the following boundary-value problems:

1. $y'' - y' - 6y = 0$, when $x = 0, y = 1$ and when $x = 1, y = 0$

Ans: $y = (e^{3x} - e^{5-2x})/(1 - e^{5})$

2. $y'' - 10y' + 25y = 0$, when $x = 0, y = 1$ and when $x = 1, y = 0$

Ans: $y = e^{5x} - xe^{5x}$

3. $y'' + y = 0$, when $x = 0, y = 0$ and when $x = \pi, y = 0$

Ans: $y = c \sin x$

e. If the roots of an auxiliary equation are $\lambda_1 = 4$, $\lambda_2 = \lambda_3 = -1$, find the corresponding differential equation.

3.5. Nonhomogeneous Linear Equations With Constant Coefficients

The different equation takes the form

$$ay''+by'+cy=f(x) \tag{15}$$

where a, b and c are constants. To solve this equation, let us assume that the solution y is the sum of two functions in x, namely u and v, then

$$y=u+v \tag{16}$$

If this is true, y must satisfy the differential equation (15). We have

$$y'=u'+v' \text{ and } y''=u''+v''$$

Substituting for y, y' and y'' in Equation (15), we get

$$a(u''+v'')+b(u'+v')+c(u+v)=f(x)$$

Or

$$(au''+bu'+cu)+(av''+bv'+cv)=f(x) \tag{17}$$

Equation (17) can be satisfied if

$$au''+bu'+cu=0 \tag{18}$$

and

$$av''+bv'+cv=f(x) \tag{19}$$

Equation (18) represents the ***related homogeneous equation*** of the original equation. Its solution u is a function of x and is called the ***complementary function*** y_{CF}. It can be found using the techniques presented in the previous subsection. On the other hand, the solution v of Equation (19) represents the ***particular integral*** or ***particular solution,*** and is denoted by y_{PI}. Particular to what? to the function $f(x)$. If $f(x)$ is changed then the particular integral will change, whereas the complementary function remains the same. The general solution can now be obtained by following these steps:

1. Obtain the complementary function y_{CF} by solving equation (18) using the method mentioned in the previous subsection.
2. obtain the particular integral y_{PI} by solving equation (19). How?
3. The general solution of Equation (15) is then $y=y_{CF}+y_{PI}$.

The question is now how to find the particular integral? To answer this question, we need some tools. In the next subsections, we present some methods to obtain the particular solutions of the differential equation.

First, we state and prove the following important theorem.

Theorem: If y_1 and y_2 are the two linearly independent solutions of the related homogeneous equation $ay''+by'+cy=0$, and if y_{PI} is any solution of the non-homogeneous equation

$ay''+by'+cy=f(x)$,

then every solution of the non-homogeneous equation is of the form $y=c_1y_1(x)+c_2y_2(x)+y_{PI}(x)$ for some choices of the constants c_1 and c_2.

Proof: Let y be any solution of the non-homogeneous equation, then

$ay''+by'+cy=f(x)$,

and since y_{PI} is also a solution, then

$ay_{PI}''+by_{PI}'+cy_{PI}=f(x)$.

Subtracting these two equations, we obtain

$a(y-y_{PI})''+b(y-y_{PI})'+c(y-y_{PI})=0.$

therefore, $(y-y_{PI})$ is a solution of the related homogeneous equation. But y_1 and y_2 form a fundamental set of solutions of this equation. Then, there must be some constants c_1 and c_2 such that

$y-y_{PI}=c_1y_1+c_2y_2$.

From which, y is of the form

$y=c_1y_1+c_2y_2+y_{PI}$. □

Note: This theorem applies equally as well to linear differential equations with variable coefficients.

3.6. The operational Method

In general, an ***operator*** transforms a given function into another function. Let "D" denote differential operator), then

$$Dy = \frac{d}{dx}y = y' \tag{20}$$

For example, $Dx^2 = 2x$ and $D\tan x = \sec^2 x$. Applying D, we have

$$D^2y = D(Dy) = Dy' = y''.$$

It can be easily shown that the differential operator D satisfies the following laws of algebra given that m and n are positive integer,

$$(D^m + D^n)v = (D^n + D^m)v$$

$$D^m D^n v = D^n D^m v = D^{m+n} v$$

$$D(u+v) = Du + Dv$$

The differential equation $ay'' + by' + cy = 0$ can be put in operator form

$$(aD^2 + bD + c)y = 0$$

We say that $(aD^2 + bD + c)$ is a ***polynomial differential operator*** of degree two. In general,

$$P(D) = a_0(x)D^n + a_1(x)D^{n-1} + \cdots + a_n(x) \tag{21}$$

is a polynomial differential operator of degree n.

The differential operator D is ***linear***. If $y_1(x)$ and $y_2(x)$ are two function and b_1 and b_2 are two constants, then

$$D[b_1y_1(x) + b_2y_2(x)] = b_1Dy_1(x) + b_2y_2(x) \tag{22}$$

Example: $(D^2 + 2D - 1)(2e^{5x} - x^2) = (D^2 + 2D - 1)2e^{5x} + (D^2 + 2D - 1)x^2$

$$= 50e^{5x} + 20e^{5x} - 2e^{5x} + 2 + 4x - x^2 = 68e^{5x} + 2 + 4x - x^2.$$

Let P_1, P_2 and P_3 be polynomial differential operators, then we have the following properties:

1. Commutative law. $\quad P_1P_2 = P_2P_1$
2. Associative law. $\quad P_1(P_2P_3) = (P_1P_2)P_3$
3. Distributive law. $\quad P_1(P_2 + P_3) = P_1P_2 + P_1P_3$

Polynomial differential operators with constant coefficients can be ***factored***, *i.e.* $\quad a_o D^n + a_1 D^{n-1} + \cdots + a_n = a_o(D - \alpha_1)(D - \alpha_1)\cdots(D - \alpha_n)$

Where $\alpha_1, \alpha_2, \cdots, \alpha_n$ are the roots of the auxiliary equation.

Since the reciprocal of a polynomial is not polynomial, then we cannot interpret $\frac{1}{a_0D^n+\cdots+a_n}$ as a polynomial different operator. Instead, we interpret it as the ***inverse operator***, that is $g(x)=\frac{1}{a_0D^n+\cdots+a_n}f(x)$ if $(a_0D_n+...+a_n)g(x)=f(x)$.

$1/D$ or simply D^{-1} is equivalent to integration; in fact it is called the ***integral operator***. For example, $D^{-1}x = \int x dx = x^2/2$. It is important to note that D^{-1} is used to find **an** integral but not **the** complete integral (no constant of integration is added). D^{-m} denotes that the integration should be carried out *m* times successively.

Next, we give the effect of the polynomial differential operator and its inverse on some elementary functions.

1. <u>The Function</u>: $e^{\lambda x}$

$$De^{\lambda x} = \lambda e^{\lambda x}$$

$$D^2e^{\lambda x} = \lambda^2 e^{\lambda x}$$

$$D^3e^{\lambda x} = \lambda^3 e^{\lambda x}$$

$$\cdot \quad \cdot \quad \cdot$$

$$D^n e^{\lambda x} = \lambda^n e^{\lambda x}$$

If *P*(*D*) is a polynomial differential operator with constant coefficients then

$$P(D)e^{\lambda x} = P(\lambda)e^{\lambda x}$$

We can see that, for the function $e^{\lambda x}$, the operator *D* is simply replaced by λ. Also, for the inverse operator, we have

$$\frac{1}{P(D)}e^{\lambda x} = \frac{1}{P(\lambda)}e^{\lambda x}, \; P(\lambda)\neq 0$$

If $P(\lambda) = 0$, then this procedure will fail, λ is a root of the auxiliary equation, and we have to find another way to evaluate $\frac{1}{P(D)}e^{\lambda x}$.

***Example* 1**: $(D^2-3D+4)e^{-2x} = [(-2)^2-3(-2)+4]e^{-2x} = 14e^{-2x}$. □

***Example* 2**: $\frac{1}{D^2-D+4}e^{3x} = \frac{1}{3^2-3+4}e^{3x} = \frac{1}{10}e^{3x}$. □

2. <u>The Functions</u>: $\sin\beta x$, $\cos\beta x$

$$D\sin\beta x = \beta\cos\beta x \qquad D\cos\beta x = -\beta\sin\beta x$$

$$D^2\sin\beta x = (-\beta^2)\sin\beta x \qquad D^2\cos\beta x = (-\beta^2)\cos\beta x$$

$$D^4\sin\beta x = (-\beta^2)^2\sin\beta x \qquad D^4\cos\beta x = (-\beta^2)^2\cos\beta x$$

$$\cdots\cdots \qquad \cdots\cdots$$

$$D^{2n}\sin\beta x = (-\beta^2)^n\sin\beta x \qquad D^{2n}\cos\beta x = (-\beta^2)^n\cos\beta x$$

If $P(D^2)$ is a polynomial differential operator with constant coefficients then

$$P(D^2)\begin{Bmatrix}\sin\beta x\\ \cos\beta x\end{Bmatrix} = P(-\beta^2)\begin{Bmatrix}\sin\beta x\\ \cos\beta x\end{Bmatrix}, \text{ and}$$

$$\frac{1}{P(D^2)}\begin{Bmatrix}\sin\beta x\\ \cos\beta x\end{Bmatrix} = \frac{1}{P(-\beta^2)}\begin{Bmatrix}\sin\beta x\\ \cos\beta x\end{Bmatrix}, \; P(-\beta^2)\neq 0.$$

If $P(-\beta^2)=0$, then this procedure will fail and we have to find another way to evaluate $\frac{1}{P(D^2)}\sin\beta x$ or $\frac{1}{P(D^2)}\cos\beta x$. We can see that, in general, for the functions $\sin\beta x$ and $\cos\beta x$, every D^2 is replaced by $-\beta^2$. Similarly, for the hyperbolic functions $\sinh kx$ and $\cosh kx$, D^2 every will be replaced by k^2.

We have
$$P(D^2)\begin{Bmatrix}\sinh kx\\ \cosh kx\end{Bmatrix} = P(k^2)\begin{Bmatrix}\sin kx\\ \cos kx\end{Bmatrix}, \text{ and}$$

$$\frac{1}{P(D^2)}\begin{Bmatrix}\sinh kx\\ \cosh kx\end{Bmatrix} = \frac{1}{P(k^2)}\begin{Bmatrix}\sinh kx\\ \cosh kx\end{Bmatrix}.$$

***Example* 1:** $\frac{1}{D^2+5}\sin 2x = \frac{1}{(-2^2)+5}\sin 2x = \sin 2x$. □

***Example* 2:** $\frac{1}{D^2+5}(e^{-2x}+\cos 3x) = \frac{1}{D^2+5}e^{-2x} + \frac{1}{D^2+5}\cos 3x$

$$= \frac{1}{(-2)^2+5}e^{-2x} + \frac{1}{-3^2+5}\cos 3x = \frac{1}{9}e^{-2x} - \frac{1}{4}\cos 3x. \; □$$

***Example* 3:** $\frac{1}{D^2+1}\cosh 2x = \frac{1}{2^2+1}\cosh 2x = \frac{1}{5}\cosh 2x$. □

Note: The problem of ***Example* 2** is equivalent to finding the particular solution of the equation $(D^2+5)y = e^{-2x}+\cos 3x$.

3. <u>Functions of the form</u> $e^{\lambda x} v(x)$ (*the* ***exponential shift***)

Here $v(x)$ is any function in the variable x. Applying the operator D, we get

$$D e^{\lambda x} v = e^{\lambda x} Dv + v \lambda e^{\lambda x} = e^{\lambda x} (D + \lambda) v$$

Successive applications of the operator D yields

$$D^2 e^{\lambda x} v = e^{\lambda x} (D + \lambda)^2 v$$

$$D^3 e^{\lambda x} v = e^{\lambda x} (D + \lambda)^3 v$$

$$\cdots\cdots$$

$$D^n e^{\lambda x} v = e^{\lambda x} (D + \lambda)^n v$$

It is clear that, in this case, every D is shifted by λ. Moreover if $P(D)$ is a polynomial differential operator, then

$$P(D) e^{\lambda x} v = e^{\lambda x} P(D + \lambda) v \text{ , and}$$

$$\frac{1}{P(D)} e^{\lambda x} v = e^{\lambda x} \frac{1}{P(D + \lambda)} v$$

***Example* 1**: Evaluate $\dfrac{1}{D^2 + 2D + 3} e^{-x} \sin 2x$.

Solution: Using the exponential shift property, we obtain

$$\frac{1}{D^2 + 2D + 3} e^{-x} \sin 2x = e^{-x} \frac{1}{(D-1)^2 + 2(D-1) + 3} \sin 2x$$

$$= e^{-x} \frac{1}{D^2 + 2} \sin 2x = e^{-x} \frac{1}{-2^2 + 2} \sin 2x = -\frac{1}{2} e^{-x} \sin 2x \ . \quad \square$$

***Example* 1**: Evaluate $\dfrac{1}{D^2 + D + 1} e^{2x} \cosh 3x$.

Solution: Using the exponential shift property, we obtain

$$\frac{1}{D^2 + D + 1} e^{2x} \cosh 3x = e^{2x} \frac{1}{(D+2)^2 + (D+2) + 1} \cosh 3x$$

$$= e^{2x} \frac{1}{D^2 + 5D + 7} \cosh 3x = e^{2x} \frac{1}{3^2 + 5D + 7} \cosh 3x$$

$$= e^{2x} \frac{1}{5D + 16} \cosh 3x = e^{2x} \frac{5D - 16}{25D^2 - 256} \cosh 3x$$

$$= e^{2x} \frac{5D - 16}{25(3^2) - 256} \cosh 3x = -\frac{1}{31} e^{2x} (5D - 16) \cosh 3x$$

$$= \frac{1}{31} e^{2x} (16 \cosh 3x - 15 \sinh 3x) \ . \quad \square$$

Note: The exponential shift can be used when λ in $e^{\lambda x}$ is also a root of the auxiliary equation $P(\lambda)=0$. The following example illustrates this particular case.

***Example* 3**: Evaluate $\dfrac{1}{D^2-3D-4}e^{4x}$.

Solution: Clearly, if we replace every D by 4, the denominator will vanish. In this case we use the exponential shift property as follows.

$$\frac{1}{D^2-3D-4}e^{4x}=e^{4x}\frac{1}{(D+4)^2-3(D+4)-4}\cdot 1$$
$$=e^{4x}\frac{1}{D^2+5D}\cdot 1=\frac{1}{5}e^{4x}\frac{1}{1+D/5}\cdot x$$
$$=\frac{1}{5}e^{4x}(1+\frac{D}{5}+\cdots)\cdot x=\frac{1}{5}e^{4x}(x-\frac{1}{5})$$

Note: 1.The integral operator $1/D$ appeared in the previous example when it operates on 1 we get x without the constant of integration. In general we have

$$\frac{1}{D^m}\cdot 1=\frac{x^m}{m!},\quad m=1,2,\cdots$$

$$\frac{1}{(D-\lambda)^m}\cdot e^{\lambda x}=\frac{x^m e^{\lambda x}}{m!},\quad m=1,2,\cdots$$

$$\frac{1}{(D-\lambda)^m P(D)}\cdot e^{\lambda x}=\frac{x^m e^{\lambda x}}{m!P(\lambda)},\quad m=1,2,\cdots,\quad P(\lambda)\neq 0$$

2. We have used the binomial expansion of the operator $\dfrac{1}{1+D/5}$.

3. In some cases, we can factor the polynomial differential operator $P(D)$. If $\lambda_1,\lambda_2,\cdots,\lambda_n$ are the roots of the auxiliary equation, then

$$P(D)=a_0D^n+a_1D^{n-1}+\cdots+a_n=a_0(D-\lambda_1)(D-\lambda_2)\cdots(D-\lambda_n)$$

Factorization can provide us with a tool to find the particular integral of an equation. Using the exponential shift, we have

$$\frac{1}{D-\lambda}\cdot f(x)=\frac{1}{D-\lambda}\cdot e^{\lambda x}e^{-\lambda x}f(x)=e^{\lambda x}\frac{1}{D+\lambda-\lambda}\cdot e^{-\lambda x}f(x)$$
$$=e^{\lambda x}\frac{1}{D}\cdot e^{-\lambda x}f(x)=e^{\lambda x}\int e^{-\lambda x}f(x)dx$$

Therefore $\dfrac{1}{D-\lambda}\cdot f(x)=e^{\lambda x}\int e^{-\lambda x}f(x)dx$ .The following examples illustrate the use of this important relation.

***Example* 4**: Find the particular integral of $(D^2 - 1)y = e^{-x}$.

Solution: Factoring the polynomial differential operator, we get

$(D+1)(D-1)y = e^{-x}$.

Let $(D-1)y = u$ then $(D+1)u = e^{-x}$. Hence

$u = e^{-x} \int e^x e^{-x}\,dx = x e^{-x}$, and

$$y = e^x \int e^{-x} u\,dx = e^x \int x e^{-2x}\,dx = e^x\left(-\frac{x e^{-2x}}{2} - \frac{e^{-2x}}{4}\right),$$

and the particular integral is $y_{PI} = -\frac{1}{2} x e^{-x} - \frac{1}{4} e^{-x}$. □

***Example* 5**: Find the particular integral of $(D^2 - 1)y = 2(\sec^3 x - \sec x)$.

Solution: Factoring the polynomial differential operator, we get

$(D+1)(D-1)y = 2(\sec^3 x - \sec x)$.

Let $(D-1)y = u$ then $(D+1)u = 2\sec^3 x - 2\sec x$. Hence

Then $u = 2e^{-x} \int e^x (\sec^3 x - \sec x)\,dx = \sec x \tan x - \sec x$, and

$y = e^x \int e^{-x} (\sec x \tan x - \sec x)\,dx = \sec x$.

The particular integral is $y_{PI} = \sec x$. □

***Example* 6**: Find the particular integral of $(D^2 - 1)y = \dfrac{2}{e^x + e^{-x}}$.

Solution: Factoring the polynomial differential operator, we get

$(D+1)(D-1)y = \dfrac{2}{e^x + e^{-x}}$.

Let $(D-1)y = u$ then $(D+1)u = \dfrac{2}{e^x + e^{-x}}$. Hence

$$u = 2e^{-x} \int \frac{e^x}{e^x + e^{-x}}\,dx = 2e^{-x} \int \frac{e^{2x}}{e^{2x} + 1}\,dx = e^{-x} \ln(e^{2x} + 1)$$

$$y = e^x \int e^{-2x} \ln(e^{2x} + 1)\,dx = x e^x - \frac{(e^x + e^{-x}) \ln(e^{2x} + 1)}{2}$$

The particular solution is $y_{PI} = x e^x - \dfrac{(e^x + e^{-x}) \ln(e^{2x} + 1)}{2}$ □

***Example* 7:** Find the particular integral of $(D^2 + 2D + 1)y = (e^x - 1)^{-2}$.

***Solution*:** Factoring the polynomial differential operator, we get

$(D+1)(D+1)y = (e^x - 1)^{-2}$.

Let $(D+1)y = u$ then $(D+1)u = (e^x - 1)^{-2}$.

Hence

$$u = e^{-x} \int \frac{e^x \, dx}{(e^x - 1)^2} = -\frac{e^{-x}}{e^x - 1}, \text{ and}$$

$$y = -e^{-x} \int \frac{dx}{e^x - 1} = x e^{-x} - e^{-x} \ln(e^x - 1).$$

The particular integral is $y_{PI} = x e^{-x} - e^{-x} \ln(e^x - 1)$. □

The following table gives a summary of the effect of the inverse operator used to find the particular solution of some linear differential equations.

***Table* 2**: Effect of the inverse operator

Expression	Value
$\frac{1}{P(D)}e^{\lambda x}$	$\frac{e^{\lambda x}}{P(\lambda)}, \quad P(\lambda) \neq 0$
$\frac{1}{(D-\lambda)^m}e^{\lambda x}$	$\frac{x^m e^{\lambda x}}{m!}, \quad m = 1, 2, \ldots$
$\frac{1}{(D-\lambda)^m P(D)}e^{\lambda x}$	$\frac{x^m e^{\lambda x}}{m!P(\lambda)}, \quad m = 1, 2, \cdots; \ P(\lambda)$
$\frac{1}{D^2+\alpha^2}\sin\beta x$	$\frac{\sin\beta x}{\alpha^2-\beta^2}, \quad \alpha \neq \beta$
$\frac{1}{D^2+\alpha^2}\cos\beta x$	$\frac{\cos\beta x}{\alpha^2-\beta^2}, \quad \alpha \neq \beta$
$\frac{1}{aD^2+bD+c}\sin\beta x$	$\frac{(c-a\beta^2)\sin\beta x - b\beta\cos\beta x}{(c-a\beta^2)^2+b^2\beta^2}$, denom $\neq 0$
$\frac{1}{aD^2+bD+c}\cos\beta x$	$\frac{(c-a\beta^2)\cos\beta x + b\beta\cos\beta x}{(c-a\beta^2)^2+b^2\beta^2}$, denom $\neq 0$
$\frac{1}{P(D)}\left[c_1 f_1(x) + c_2 f_2(x)\right]$	$c_1\frac{1}{P(D)}f_1(x) + c_2\frac{1}{P(D)}f_2(x)$
$\frac{1}{P_1(D)P_2(D)}f(x)$	$\frac{1}{P_1(D)}\left[\frac{1}{P_2(D)}f(x)\right]$
$\frac{1}{D}f(x)$	$\int f(x)dx$
$\frac{1}{D-a}f(x)$	$e^{ax}\int e^{-ax}f(x)dx$
$\frac{1}{(D-a)^m}f(x)$	$\frac{e^{ax}}{(m-1)!}\int_0^x e^{-au}(x-u)^{m-1}f(u)du$
$\frac{1}{P(D)}e^{\lambda x}f(x)$	$e^{\lambda x}\frac{1}{P(D+\lambda)}f(x)$

The knowledge gained in using the polynomial differential operator and its inverse can help us find the particular solution of some linear differential equations with constant coefficients. The following examples illustrate the idea.

***Example* 1**: Find the general solution of $(D^2 - 7D + 12)y = e^{5x}$.

Solution: The auxiliary equation is $\lambda^2 - 7\lambda + 12 = 0$, and the roots are $\lambda_1 = 4$ and $\lambda_2 = 3$.

The complementary function is $y_{CF} = Ae^{4x} + Be^{3x}$.

For the particular integral, we have

$$y_{PI} = \frac{1}{D^2 - 7D + 12} e^{5x} = \frac{1}{25 - 37 + 12} e^{5x} = \frac{1}{2} e^{5x}.$$

Then, the general solution is

$$y = y_{CF} + y_{PI} = Ae^{4x} + Be^{3x} + \frac{1}{2} e^{5x}$$ □

Note: Remember that the complementary function is a linear combination of the two independent solutions of the related homogeneous equation, and contains two arbitrary constants. Whereas the particular integral does not contain any constant.

***Example* 2**: Solve the equation $(D^2 - 3D + 2)y = x^2$.

Solution: The auxiliary equation is $\lambda^2 - 3\lambda + 2 = 0$, and the roots are $\lambda_1 = 2$ and $\lambda_2 = 1$.

The complementary function will be $y_{CF} = Ae^{2x} + Be^{x}$.

For the particular integral, we have

$$y_{PI} = \frac{1}{(D-2)(D-1)} x^2 = \left\{ \frac{1}{D-2} - \frac{1}{D-1} \right\} x^2 \text{ (partial fractions)}$$

$$= \left\{ -\frac{1}{2} \frac{1}{1 - D/2} + \frac{1}{1-D} \right\} x^2 \text{ (using the binomial expansion)}$$

$$= \left\{ -\frac{1}{2} \left(1 + \frac{D}{2} + \frac{D^2}{4} + \cdots \right) + \left(1 + D + D^2 + \cdots \right) \right\} x^2$$

$$= -\frac{1}{2} \left(x^2 + x + \frac{1}{2} \right) + \left(x^2 + 2x + 2 \right) = \frac{1}{2} x^2 + \frac{3}{2} x + \frac{7}{4}$$

Now, the general solution is $y = Ae^{2x} + Be^{x} + \frac{1}{2} x^2 + \frac{3}{2} x + \frac{7}{4}$. □

***Example* 3**: Solve the differential equation $(D^2 - 6D + 9)y = 2x\, e^{3x}$.

Solution: The auxiliary equation is $\lambda^2 - 9\lambda + 6 = 0$, and the roots are $\lambda_1 = \lambda_2 = 3$.

The complementary function will be $y_{CF} = (Ax + B)e^{3x}$.

For the particular integral, we write $e^{-3x}(D-3)^2 y_{PI} = 2x$.

Using the shifting property, we obtain $D^2\left[e^{-3x} y_{PI}\right] = 2x$.

Integrating twice, we get $e^{-3x} y_{PI} = \frac{1}{3}x^3$, the general solution is

$$y = (Ax + b)e^{3x} + \frac{1}{3}x^3 e^{3x} .$$

□

***Example* 4**: Solve the differential equation $(D^2 + 6D + 9)y = -625\cos 4x$.

Solution: The auxiliary equation is $\lambda^2 + 9\lambda + 6 = 0$, and we have a double root at $\lambda = -3$. The complementary function will be

$y_{CF} = (A + Bx)e^{-3x}$.

For the particular integral, we have.

$$y_{PI} = \frac{1}{D^2 + 6D + 9}(-625\cos 4x) = -625\frac{1}{-16 + 6D + 9}\cos 4x$$

$$= -625\frac{1}{6D - 7}\cos 4x = -625\frac{6D + 7}{36D^2 - 49}\cos 4x$$

$$= -625\frac{6D + 7}{-576 - 49}\cos 4x = (6D + 7)\cos 4x = 7\cos 4x - 24\sin 4x$$

The general solution is $y = (A + Bx)e^{-3x} + 7\cos 4x - 24\sin 4x$. □

***Example* 5**: Solve the differential equation $(D^3 + 4D)y = \sin 3x$.

Solution: The auxiliary equation is $\lambda^3 + 4\lambda = 0$, and the roots are $\lambda_1 = 0$, $\lambda_2 = 2i$ and $\lambda_3 = -2i$. The complementary function will be

$y_{CF} = A + B\cos 2x + C\sin 2x$.

For the particular integral, we have

$$y_{PI} = \frac{1}{D(D^2 + 4)}\sin 3x = \frac{1}{D(-3^2 + 4)}\sin 3x = -\frac{1}{5}\cdot\frac{1}{D}\sin 3x$$

$$= -\frac{1}{5}\int \sin 3x\, dx = \frac{1}{15}\cos 3x ,$$

and the general solution is $y = A + B\cos 2x + C\sin 2x + \frac{1}{15}\cos 3x$ □

***Example* 6**: Solve the equation $(D-2)(D^2+3D+2)y = e^{2x}\cos 3x$.

Solution: The auxiliary equation is $(\lambda-2)(\lambda^2+3\lambda+2)=0$, and the roots are $\lambda_1 = 2$, $\lambda_2 = -2$ and $\lambda_3 = -1$. The complementary function is $y_{CF} = Ae^{2x} + Be^{-2x} + Ce^{-x}$.

For the particular integral, we have

$$y_{PI} = \frac{1}{(D-2)(D^2+3D+2)} e^{2x}\cos 3x .$$

Using the exponential shift property, we get

$$y_{PI} = e^{2x} \frac{1}{(D+2-2)[(D+2)^2+3(D+2)+2]}\cos 3x$$

$$= e^{2x}\frac{1}{D(D^2+7D+12)}\cos 3x = \frac{1}{3}e^{2x}\frac{1}{-3^2+7D+12}\sin 3x$$

$$= \frac{1}{3}e^{2x}\frac{1}{7D+3}\sin 3x = \frac{1}{3}e^{2x}\frac{7D-3}{49D^2-9}\sin 3x$$

$$= -\frac{1}{1350}e^{2x}(7D-3)\sin 3x = \frac{1}{450}e^{2x}(\sin 3x - 7\cos 3x)$$

The general solution is

$$y = Ae^{2x} + Be^{-2x} + Ce^{-x} + \frac{1}{450}e^{2x}(\sin 3x - 7\cos 3x).$$ □

***Example* 7**: Solve the equation $(D-1)^2 y = e^x \sec^2 x \tan x$.

Solution: The auxiliary equation is $(\lambda-1)^2 = 0$, then we have a double root at $\lambda = 1$. The complementary function will be $y_{CF} = (A+Bx)e^x$.

Now,

$$y_{PI} = \frac{1}{(D-1)^2}e^x \sec^2 x \tan x = e^x \frac{1}{D^2}\sec^2 x \tan x$$

$$= e^x \frac{1}{D}\int \sec^2 x \tan x\, dx = \frac{1}{2}e^x \frac{1}{D}\tan^2 x = \frac{1}{2}e^x(\tan x - x).$$

But the term $\frac{1}{2}xe^x$ is a part of the complementary function, then the particular solution becomes $y_{PI} = \frac{1}{2}e^x \tan x$

and the general solution is $y = (A+Bx)e^x + \frac{1}{2}e^x \tan x$. □

***Example* 8**: Solve the equation $(D^2+a^2)y=\sin ax$.

Solution: The auxiliary equation is $\lambda^2+a^2=0$, and $\lambda=\pm i\,a$, then the complementary function is $y_{CF}=A\cos ax+B\sin ax$.

Now, we can see that a is the imaginary part of one of the roots of the auxiliary equation, then we cannot, in this case, simply use the substitution of $-a^2$ for D^2 in $\dfrac{1}{D^2+a^2}$. But we still can use the definition of the function $\sin ax$ in terms of $e^{\pm i\,ax}$ given by $\sin ax=\dfrac{e^{i\,ax}-e^{-i\,ax}}{2i}$, and then use the exponential shift property.

We have

$$y_{PI}=\frac{1}{D^2+a^2}\sin ax=\frac{1}{(D+ia)(D-ia)}\left[\frac{e^{iax}-e^{-iax}}{2i}\right]$$

$$=\frac{1}{4a}\left[\frac{1}{D+ia}-\frac{1}{D-ia}\right]\left(e^{iax}-e^{-iax}\right)\ \text{(partial fractions)}$$

$$=\frac{1}{4a}\left[\frac{1}{a}\frac{e^{iax}-e^{-iax}}{2i}-\left(e^{iax}+e^{-iax}\right)\frac{1}{D}\cdot 1\right]$$

$$=\frac{1}{4a^2}\sin ax-\frac{1}{2a}x\cos ax .$$

But the term $\dfrac{1}{4a^2}\sin ax$ is a part of the complementary function, then the particular integral reduces to $y_{PI}=-\dfrac{1}{2a}\cos ax$, and the general solution is $y=A\cos ax+B\sin ax-\dfrac{1}{2a}x\cos ax$. □

Note: 1. From the previous example, we have

$$\frac{1}{D^2+a^2}\sin ax=-\frac{1}{2a}x\cos ax \text{ and } \frac{1}{D^2+a^2}\cos ax=\frac{1}{2a}x\sin ax .$$

2. C.A. Hutchinson in the Amer. Math. Monthly (from 1933 to 1937), gave some notes on operational formulae. All these formulae are based on the definitions of trigonometric and hyperbolic functions in terms of exponential functions. We state here some of these formulae.

$$\frac{1}{P(D)}\sin\beta x = \frac{P(-D)\sin\beta x}{P(i\beta)P(-i\beta)};\ P(i\beta)P(-i\beta)\neq 0$$

$$\frac{1}{P(D)}\cos\beta x = \frac{P(-D)\cos\beta x}{P(i\beta)P(-i\beta)};\quad P(i\beta)P(-i\beta)\neq 0$$

$$\frac{1}{P(D)}\sinh kx = \frac{P(-D)\sinh kx}{P(k)P(-k)};\ P(k)P(-k)\neq 0$$

$$\frac{1}{P(D)}\cosh kx = \frac{P(-D)\cosh kx}{P(k)P(-k)};\quad P(k)P(-k)\neq 0$$

$$\frac{1}{(D^2+\beta^2)^n}\sin\beta x = \frac{x^n}{(2\beta)^n n!}\sin(\beta x - \tfrac{1}{2}n\pi)$$

$$\frac{1}{(D^2+\beta^2)^n}\cos\beta x = \frac{x^n}{(2\beta)^n n!}\cos(\beta x - \tfrac{1}{2}n\pi)$$

Exercise 3.4

a. Find the general solution of the following differential equations:

1. $(D^2+D)y = -\cos x$ **Ans:** $y = A + Be^{-x} + \frac{1}{2}\cos x - \frac{1}{2}\sin x$
2. $(D^2-6D+9)y = e^x$ **Ans:** $y = (a+bx)e^{3x} + \frac{1}{4}e^x$
3. $(D^2+3D+2)y = 12x^2$ **Ans:** $y = Ae^{-x} + Be^{-2x} + 6x^2 - 18x + 21$
4. $(D^2-4)y = e^{2x} + 2$ **Ans:** $y = ae^{-2x} + (b + \frac{1}{4}x)e^{2x} - \frac{1}{2}$
5. $(D^2+1)y = \cos x$ **Ans:** $y = a\cos x + b\sin x + \frac{1}{2}x\sin x$
6. $(D^3-D^2+D-1)y = 4\sin x$ **Ans:** $y = ae^x + (b+x)\cos x + (c-x)\sin x$
7. $(D^4-1)y = e^{-x}$ **Ans:** $y = ae^x + (b - \frac{1}{4}x)e^{-x} + c\cos x + d\sin x$
8. $(D^2-1)y = 10\sin^2 x$ **Ans:** $y = ae^x + be^{-x} - 5 + \cos 2x$
9. $(D^2+1)y = 12\cos^2 x$ **Ans:** $y = a\cos x + b\sin x + 6 - 2\cos 2x$
10. $(D^2+4)y = 8x + 1 - 15e^x$ **Ans:** $y = a\cos 2x + b\sin 2x + 2x + \frac{1}{4} - 3e^x$
11. $(D+2)^2 y = 12x\,e^{-2x}$ **Ans:** $y = (a + bx + 2x^3)e^{-2x}$
12. $(D^2+2D+1)y = 48e^{-x}\cos 4x$ **Ans:** $y = (a + bx - 3\cos 4x)e^{-x}$
13. $(D^2+4D+4)y = -x^{-2}e^{-2x}$ **Ans:** $y = e^{-2x}(a + bx + \ln x)$

14. $(D-a)^2 y = e^{ax} f''(x)$ **Ans:** $y = e^{ax}[a+bx+f(x)]$

15. $(D^2+7D+12)y = e^{-3x}\sec^2 x(1+2\tan x)$

Ans: $y = ae^{-4x} + e^{-3x}(b+\tan x)$

16. $(D^2-3D+2)y = 2x^3-9x^2+2x-16$

Ans: $y = ae^{2x} + be^{x} + x^3 - 2x - 11$

17. $(D^2+16)y = 24\sin 4x$ **Ans:** $y = (a-3x)\cos 4x + b\sin 4x$

18. $(D^2-2D+5)y = e^{x}\cos 2x$

Ans: $y = e^{x}(a\cos 2x + b\sin 2x) + \frac{1}{4}x\, e^{x}\sin 2x$

19. $(D^6-1)y = x^{10}$ **Ans:** $y_{PI} = -x^{10} - \frac{10!}{4!}x^4$

20. $(D^2+4)y = 20(e^{x} - \cos 2x)$ **Ans:** $y = a\cos 2x + (b-5x)\sin 2x + 4e^{x}$

21. $D(D^2+1)y = \sin x$ **Ans:** $y = a + b\cos x + (c - \frac{1}{2}x)\sin x$

22. $D^2(D^2+1)y = \sin x$ **Ans:** $y = a + bx + (c + \frac{1}{2}x)\cos x + d\sin x$

23. $(D^4-1)y = 5\cos x\cosh x$

Ans: $y = a\cosh x + b\sinh x + c\cos x + d\sin x - \cosh x\cos x$

24. $(4D^2+4D+1)y = 4\sin\frac{x}{2}$ **Ans:** $y = e^{-x/2}(a+bx) - 2\cos\frac{x}{2}$

25. $(D-1)^2 y = 2e^{x}$ **Ans:** $y = (x^2 + ax + b)\,e^{x}$

26. $(D^2-2D-3)y = 12e^{5x}$ **Ans:** $y = ae^{3x} + be^{-x} + e^{5x}$

27. $(D^4+5D^2+4)y = 40\sin 3x$

Ans: $y = a\cos 2x + b\sin 2x + c\cos x + d\sin x + \sin 3x$

28. $(D^2+a^2)y = \tan ax$

Ans: $y = c_1\cos ax + c_2\sin ax - \frac{1}{a^2}\cos ax\,\ln\tan[(\pi+2ax)/4]$

29. $(D^2-3D+2)y = \sec^2 e^{-x}$ **Ans:** $y = ae^{x} + be^{2x} + e^{2x}\ln\sec e^{-x}$

30. $(D^3+D)y = \sec^2 x$

Ans: $y = a + b\cos x + c\sin x - \cos x\,\ln(\sec x + \tan x)$

31. $(D^2+2D+1)y=(e^x+1)^{-2}$ **Ans:** $y=e^{-x}[a+bx+\ln(1+e^{-x})]$

32. $(D^2+4D+3)y=\sin e^x$

Ans: $y=ae^{-x}+be^{-3x}-e^{-2x}\sin e^x-e^{-3x}\cos e^x$

b. Solve the following initial value problems:

1. $(D^2+1)y=10e^{2x}$; when $x=0, y=0$ and $y'=0$

Ans: $y=2(e^{2x}-\cos x-2\sin x)$

2. $(D^2-4)y=2-8x$; when $x=0, y=0$ and $y'=5$

Ans: $y=e^{2x}-\frac{1}{2}e^{-2x}+2x-\frac{1}{2}$

3. $(D^2+3D)y=-18x$; when $x=0, y=0$ and $y'=5$

Ans: $y=1+2x-3x^2-e^{-3x}$

4. $\frac{d^2x}{dt^2}+4\frac{dx}{dt}+5x=10$; when $t=0, x=0$ and $\frac{dx}{dt}=0$

Ans: $x=2(1-e^{-2t}\cos t-2e^{-2t}\sin t)$

5. $\frac{d^2x}{dt^2}+4\frac{dx}{dt}+5x=8\sin t$; when $t=0, x=0$ and $\frac{dx}{dt}=0$

Ans: $x=(1+e^{-2t})\sin t-(1-e^{-2t})\cos t$

6. $(D^2+4)y=2x-8$; when $x=0, y=1$ and $y'=0$

Ans: $4y=12\cos 2x-\sin 2x+2x-8$

7. $(D^2-2D+5)y=e^x\cos^2 x$; when $x=0, y=0$ and $y'=1$

Ans: $y=\frac{1}{8}e^x(4\sin 2x-\cos 2x+x\sin 2x+1)$

8. $(D^2-1)y=\sin 2x$; when $x=0, y=0$ and $y'=1$

Ans: $5y=7\sinh x-\sin 2x$

c. For the equation $(D^3+D^2)y=4$, find the solution whose curve has at the origin a point of inflection with a horizontal tangent line.

Ans: $y=4-4x-2x^2-4e^{-x}$

d. Solve the boundary value problem $(D^2+1)y=2\cos x$; when $x=0$, $y=0$, and when $x=\pi$, $y=0$. Show that the boundary value problem has an unlimited number of solutions. **Ans:** $y=(c+x)\sin x$

e. Show that the boundary value problem $(D^2+1)y = x^3$; when $x = 0$, $y = 0$, and when $x = \pi$, $y = 0$ has no solution.

f. Use the substitution $x = \tan\theta$ to transform the differential equation:

$$(1+x^2)^3 \frac{d^2y}{dx^2} + 2x(1+x^2)^2 \frac{dy}{dx} + (1+x^2)y = 3x$$

into a differential equation with constant coefficients. Then find the solution of the original equation when $x = 0, y = 0$ and $y' = 0$.

Ans: $\dfrac{d^2y}{d\theta^2} + y = 3\sin\theta\cos\theta,\ y = \dfrac{x}{1+x^2}\left[\sqrt{1+x^2} - 1\right]$

g. Solve the equation: $4x\dfrac{d^2y}{dx^2} + 2(1-\sqrt{x})\dfrac{dy}{dx} - 6y = e^{-2\sqrt{x}}$, using the substitution $u = \sqrt{x}$. **Ans:** $y = ae^{3\sqrt{x}} + be^{-2\sqrt{x}} - \frac{1}{5}\sqrt{x}\,e^{-2\sqrt{x}}$

h. Solve the equation: $\cos x\dfrac{d^2y}{dx^2} + \sin x\dfrac{dy}{dx} + 4y\cos^3 x = 8\cos^5 x$, using the substitution $t = \sin x$. **Ans:** $y = a\sin(2\sin x) + b\cos(2\sin x) + 3 - 2\sin^2 x$

3.7. The Method of Undetermined Coefficients

Consider again the linear second order differential equation with constant coefficients

$$(aD^2 + bD + c)y = f(x) \tag{23}$$

Although the method of undetermined coefficients is not limited to second order differential equations, it is limited to non-homogeneous linear equations with constant coefficients, and where $f(x)$ is a constant, polynomial function, exponential function, sine/cosine functions or finite sums and products of these functions.

The following are some examples of the types of such input functions $f(x)$ that are appropriate for this discussion:

$f(x) = 10$ (a constant)

$f(x) = x^2 - 5x$ (a polynomial function)

$f(x) = 15x - 6 + 8e^{-4x}$ (polynomial + exponential)

$f(x) = \sin 3x - 5x \cos 2x$

$f(x) = e^x \cos x - (3x^2 - 1)e^{-x}$

....

The method of undetermined coefficients is not applicable to equations of the type (23) when $f(x) = \ln x, \tan x, \sin^{-1} x, \cdots$. Differential equations with the latter kind of inputs will be considered later on.

Basic idea:

The set of functions {constants, polynomials, exponentials, sines and cosines} has the remarkable property that derivatives of their sums and products are again sums and products of {constants, polynomials, exponentials, sines and cosines}.

Since the linear combination of derivatives $a y''_{PI} + b y'_{PI} + c y_{PI}$ must be identically $f(x)$, it seems reasonable to assume that y_{PI} has the same form as $f(x)$ except some exceptional cases.

The following examples will illustrate the basic idea.

***Example* 1:** Solve the differential equation $(D^2 - 3D + 2)y = x^2$.

***Solution*:** We first solve the related homogeneous equation. The auxiliary equation is $\lambda^2 - 3\lambda + 2 = 0$ and the roots are $\lambda_1 = 1$ and $\lambda_2 = 2$. Hence the complementary function is $y_{PI} = c_1 e^x + c_2 e^{2x}$.

Now, since $f(x)$ is a quadratic polynomial, assume a particular solution in the form $y = a_0x^2 + a_1x + a_2$.

We seek to determine the coefficients a_0, a_1 and a_2. Differentiating gives us $y' = 2a_0x + a_1$ and $y'' = 2a_0$, and substituting into the given differential equation produces

$$y'' - 3y' + 2y = 2a_0 - 3(2a_0x + a_1) + 2(a_0x^2 + a_1x + a_2) \equiv x^2.$$

Since the last equation is supposed to be identity, the coefficients of like powers of x must be equal:

coefficients of x^2: $2a_0 = 1$

coefficients of x^1: $-6a_0 + 2a_1 = 0$

coefficients of x^0: $2a_0 - 3a_1 + 2a_2 = 0$

Hence $a_0 = \frac{1}{2}$, $a_1 = \frac{3}{2}$ and $a_2 = \frac{7}{4}$.

Thus, the particular solution is $y_{PI} = \frac{1}{2}x^2 + \frac{3}{2}x + \frac{7}{4}$,

and the general solution is $y = c_1e^x + c_2e^{2x} + \frac{1}{2}x^2 + \frac{3}{2}x + \frac{7}{4}$. □

***Example* 2**: Solve the differential equation $(D^2 - 3D + 2)y = 3e^x$.

Solution: The complementary function is as found in ***Example* 1**:

$$y_{PI} = c_1e^x + c_2e^{2x}.$$

Assume a particular solution of the form $y = a_0e^x$. Differentiating and substituting in the differential equation produces $0 = 3e^x$!!?. We have obviously made the wrong guess for the particular solution. The difficulty here is simply this: The term a_0e^x is already present in the complementary function. This means that any constant multiple of e^x, when substituted in the differential equation, must produce zero. What then should the form of y_{PI} be? We will see that in details later on. Let us see whether we can find a particular solution of the form $y = a_0x\,e^x$.

Using $y' = a_0x\,e^x + a_0e^x$ and $y'' = a_0x\,e^x + 2a_0e^x$, we obtain

$$\begin{aligned} y'' - 3y' + 2y &= (a_0x\,e^x + 2a_0e^x) - 3(a_0x\,e^x + a_0e^x) + 2a_0e^x \\ &\equiv 3e^x \end{aligned}$$

Since the value of a_0 is now readily determined, $a_0 = -3$, we have found a particular solution of the equation $y_{PI} = -3x e^x$. The reader is encouraged to verify by substitution that this is a particular solution of the differential equation. □

The difference between the procedures used in ***Example* 1** and that in ***Example* 2** suggests that we consider two cases. The first case reflects the situation of ***Example* 1**, i.e. *no function in the assumed particular solution duplicates a function in the complementary function*; while the second reflects that of ***Example* 2**, i.e. *a function in the assumed particular integral duplicates a function in the complementary function.*

<u>Case I</u>: No function in the assumed particular solution duplicates a function in the complementary function:

In the following table you will find some specific examples of $f(x)$ in the differential Equation (23) along with the corresponding proper form of the particular solution y_{PI}.

***Table* 3.** Form of the particular integral

	$f(x)$	Form of y_{PI}
1	10 (any constant)	A
2	$3x-2$	$Ax+B$
3	$5x^2+1$	Ax^2+Bx+C
4	x^3+x-1	Ax^3+Bx^2+Cx+D
5	$\sin 2x$ or $\cos 2x$	$A\sin 2x+B\cos 2x$
6	$3\sin 2x-5\cos 2x$	$A\sin 2x+B\cos 2x$
7	e^{3x}	Ae^{3x}
8	$4e^{3x-1}$	Ae^{3x}
9	$(3x+1)e^{3x}$	$(Ax+B)e^{3x}$
10	x^2e^{-3x}	$(Ax^2+Bx+C)e^{-3x}$
11	$e^{3x}\sin 2x$	$e^{3x}(A\sin 2x+B\cos 2x)$
12	$3x^2\sin 2x$	$(Ax^2+Bx+C)\sin 2x+(Dx^2+Ex+F)\cos 2x$
13	$xe^{2x}\cos 3x$	$(Ax+B)e^{2x}\cos 3x+(Cx+D)e^{2x}\sin 3x$

***Example* 1**: Determine the form of a particular integral of

i) $(D^2-8D+25)y = 5x^3e^{-x} - 7e^{-x}$

ii) $(D^2+4)y = x\cos x$

Solution: i) The auxiliary equation corresponding to the related homogeneous equation is $\lambda^2-8\lambda+25=0$, its roots are $\lambda_{1,2}=4\pm i3$.

Hence the complementary function is

$y_{CF} = c_1e^{4x}\cos 3x + c_2e^{4x}\sin 3x$.

Since $f(x)=e^{-x}(5x^3-7)$, the proper form of the particular integral is $y_{PI} = (A_0x^3+A_1x^2+A_2x+A_3)e^{-x}$.

Note that there is no duplication between the terms in the particular integral and the terms in the complementary function.

ii) The auxiliary equation corresponding to the related homogeneous equation is $\lambda^2+4=0$, its roots are $\lambda_{1,2}=\pm i2$. Hence

$y_{PI} = c_1\cos 2x + c_2\sin 2x$.

Since $f(x)=x\cos x$, the proper form for the particular integral is $y_{PI} = (A_0x+A_1)\cos x + (A_2x+A_3)\sin x$.

Again observe that there is no duplication of terms between the particular integral and the complementary function. □

***Example* 2**: Determine the form of the particular integral of

$$y''-6y'+9y = 3x^2+1+5\sin 2x - 2xe^{5x} .$$

Solution: The auxiliary equation is $\lambda^2-6\lambda+9=0$, and the roots are $\lambda=3,3$. Hence the complementary function is $y_{CF}=c_1e^{3x}+c_2xe^{3x}$.

Now, $f(x)=3x^2+1+5\sin 2x - 2xe^{5x}$, then the proper form for the particular integral is

$y_{PI} = (ax^2+bx+c)+(d\sin 2x + e\cos 2x)+(fx+g)e^{5x}$.

No term in the assumption of y_{PI} duplicates a term in y_{CF} .

***Example* 3**: Find the particular solution of $(D^2-3D-4)y = 10e^{2x}$.

Solution: Assume a particular solution of the form $y=a_0e^{2x}$, then $y'=2a_0e^{2x}$ and $y''=4a_0e^{2x}$. Substituting in the differential equation, we get $(4a_0-6a_0-4a_0)e^{2x} \equiv 10e^{2x}$. Equating the coefficients of e^{2x} in both sides, we get $a_0=-5/3$. The particular solution is $y_{PI} = -\frac{5}{3}e^{2x}$. □

***Example* 4**: Find the particular integral for the differential equation:

$$(D^2+2D+2)y=\sin 3x \ .$$

Solution: Assume a particular integral of the form $y=a_0\cos 3x+a_1\sin 3x$. Then,

$y'=-3a_0\sin 3x+3a_1\cos 3x$ and $y''=-9a_0\cos 3x-9a_1\sin 3x$.

Substituting in the differential equation, we get

$(6a_1-7a_0)\cos 3x-(6a_0+7a_1)\sin 3x\equiv\sin 3x$.

Equating the coefficients in both sides, we obtain

$6a_1-7a_0=0$ and $-6a_0-7a_1=1$, from which $a_0=-\frac{6}{85}$ and $a_1=-\frac{7}{85}$, and the particular integral will be

$$y_{PI}=-\frac{1}{85}(6\cos 3x+7\sin 3x).$$ □

***Example* 5**: Solve the differential equation: $y''+y'-2y=2x-40\cos 2x$.

Solution: The auxiliary equation is $\lambda^2+\lambda-2=0$ and the roots are $\lambda_1=1$ and $\lambda_2=-2$. The complementary function is $y_{PI}=ae^x+be^{-2x}$.

For the particular integral, we assume a solution of the form

$y=a_0x+a_1+a_2\cos 2x+a_3\sin 2x$, then

$y'=a_0-2a_2\sin 2x+2a_3\cos 2x$, and

$y''=-4a_2\cos 2x-4a_3\sin 2x$.

Substituting in the differential equation, we obtain

$$-2a_0x+(a_0-2a_1)+(-6a_2+2a_3)\cos 2x+(-6a_3-2a_2)\sin 2x$$

$$\equiv 2x-40\cos 2x$$

Equating appropriate coefficients in both sides, we get,

$a_0=-1,\quad a_1=-\frac{1}{2},\quad a_2=6$ and $a_3=-2$.

The particular integral will be $y_{PI}=-x-\frac{1}{2}+6\cos 2x-2\sin 2x$.

The general solution is

$$y=ae^x+be^{-2x}-x-\frac{1}{2}+6\cos 2x-2\sin 2x \ .$$ □

Case II: A function in the assumed particular integral duplicates a function in the complementary function:

The basic idea to avoid the duplication comes from the same idea in the case of repeated roots in the auxiliary equation of the related homogeneous equation, in which we multiply the second function by x. If the root is repeated three times, we multiply the third function by x^2, and so on. In the case of y_{PI}, we multiply the duplicated term by x. The following examples illustrate the idea.

***Example* 1**: Solve the differential equation $(D^2+1)y = 4x + 10\sin x$.

Solution: The auxiliary equation is $\lambda^2 + 1 = 0$ and the roots are $\lambda_{1,2} = \pm i$. Hence $y_{CF} = c_1 \cos x + c_2 \sin x$.

Now, $f(x) = 4x + 10\sin x$. Clearly the term $\sin x$ is present in both y_{CF} and $f(x)$. Then the proper form for the particular integral is

$$y_{PI} = ax + b + \underset{\Uparrow}{x}(c\cos x + d\sin x)$$

Differentiating this expression twice and substituting the results in the differential equation, we obtain

$$ax + b - 2c\sin x + 2d\cos x \equiv 4x + 10\sin x$$

Therefore $a = 4, b = 0, c = -5$ and $d = 0$. The particular integral is $y_{PI} = 4x - 5x\cos x$, and the general solution is

$$y = c_1\cos x + c_2\sin x + 4x - 5x\cos x \ .$$ □

***Example* 2**: Find the particular integral for the equation : $(D^2-1)y = x\,e^x$.

Solution: It is clear that $\lambda = 1$ is a root of the auxiliary equation and appears in the exponential term in $f(x)$. Then, we assume that the particular integral is of the form $y = x\,e^x(ax+b)$. Differentiating, we get

$$y' = [ax^2 + (2a+b)x + b]e^x \ , \text{ and}$$

$$y'' = [ax^2 + (4a+b)x + 2(a+b)]e^x \ .$$

Substituting in the equation, we get $(4ax + 2a + 2b)e^x \equiv x\,e^x$.

Equating corresponding coefficients we obtain $a = \frac{1}{4}$ and $b = -\frac{1}{4}$, and the particular integral is $y_{PI} = x\,e^x\left(\frac{1}{4}x - \frac{1}{4}\right)$. □

Note: If λ is a double root and if $e^{\lambda x}$ appears in $f(x)$ then it is fair to assume that the particular integral takes the form $y = ax^2 e^{\lambda x}$. In general, if λ is an n repeated root and if $e^{\lambda x}$ appears in $f(x)$ then the proper form of the particular integral is $y = ax^n e^{\lambda x}$.

***Example* 3**: Find the particular integral of $(D^2 + 1)y = \cos x$.

Solution: It is clear that $\beta = 1$ is the imaginary part of one of the roots of the auxiliary equation; then the proper form for the particular integral is

$y = x(a\cos x + b\sin x)$, $y' = (a + bx)\cos x + (b - ax)\sin x$ and

$y'' = -(2a + bx)\sin x + (2b - ax)\cos x$.

Substituting in the differential equation, we get

$-2a\sin x + 2b\cos x \equiv \cos x$.

Equating appropriate coefficients in both sides, we obtain $a = 0$ and $b = \frac{1}{2}$. Hence the particular integral will be $y_{PI} = \frac{1}{2}x\sin x$. □

***Example* 4**: Find y that satisfies the equation $D(D^2 - 1)y = 4e^{-x} + 3e^{2x}$ with the condition when $x = 0, y = 0, y' = -1$ and $y'' = 2$.

Solution: The auxiliary equation is $\lambda(\lambda^2 - 1) = 0$, and the roots are $\lambda_1 = 0$, $\lambda_2 = 1$ and $\lambda_3 = -1$. The complementary function is

$y_{CF} = A + Be^x + Ce^{-x}$.

For the particular integral, we can see that the term $4e^{-x}$ in $f(x)$ corresponds to $\lambda = -1$, and appears in y_{CF}. Then the proper form of the particular integral is $y = axe^{-x} + be^{2x}$, then

$y' = -axe^{-x} + ae^{-x} + 2be^{2x}$, $y'' = axe^{-x} - 2ae^{-x} + 4be^{2x}$,

$y''' = -axe^{-x} + 3ae^{-x} + 8be^{2x}$.

Substituting in the differential equation, we get

$2ae^{-x} + 6be^{2x} \equiv 4e^{-x} + 3e^{2x}$.

Equating coefficients in both sides, we get $a = 2$ and $b = \frac{1}{2}$.

The particular integral will be $y_{PI} = 2xe^{-x} + \frac{1}{2}e^{2x}$.

The general solution is $y = A + Be^x + (C + 2x)e^{-x} + \frac{1}{2}e^{2x}$.

To satisfy the given conditions, we have

$y' = Be^{x} - (C - 2 + 2x)e^{-x} + e^{2x}$, and

$y'' = Be^{x} + (C - 4 + 2x)e^{-x} + 2e^{2x}$

Then, at $x = 0$:

$0 = A + B + C + \frac{1}{2}$, $-1 = B - C + 3$ and $2 = B + C - 2$.

From which

$A = -\frac{9}{2}$, $B = 0$ and $C = -4$.

Then the solution satisfying the given conditions will be

$y = -\frac{9}{2} + 4e^{-x} + 2xe^{-x} + \frac{1}{2}e^{2x}$. □

***Example* 5**: Solve the differential equation $y'' - 6y' + 9y = 6x^2 + 2 - 12e^{3x}$.

Solution: The complementary function is $y_{CF} = ae^{3x} + bxe^{3x}$. The proper form for the particular integral is

$$y_{PI} = A_0x^2 + A_1x + A_2 + \underset{\Uparrow}{A_3x^3}e^{3x}$$

Differentiating twice and substituting in the differential equation we obtain

$$9A_0x^2 + (-12A_0 + 9A_1)x + 2A_0 - 6A_1 + 9A_2 + 2A_3e^{3x}$$

$$\equiv 6x^2 + 2 - 12e^{3x}$$

It follows from this identity that :

$A_0 = \frac{2}{3}, A_1 = \frac{8}{9}, A_2 = \frac{2}{3}$ and $A_3 = -6$.

Hence the general solution is

$y = ae^{3x} + bxe^{3x} + \frac{2}{3}x^2 + \frac{8}{9}x + \frac{2}{3} - 6x^2e^{3x}$. □

Note: Most of the problems in the previous section on operational method can be solved using the method of undetermined coefficients.

Exercise 3.5

a. Find the particular solution of the following differential equations using the method of undetermined coefficients:

1. $y''-4y'+3y=10e^{-2x}$ **Ans:** $y_{PI}=\frac{2}{3}e^{-2x}$
2. $y''+4y=8x^2$ **Ans:** $y_{PI}=2x^2-1$
3. $y''+y'-2y=-10\cos x$ **Ans:** $y_{PI}=3\cos x+\sin x$
4. $y''-2y'+y=e^x+x$ **Ans:** $y_{PI}=\frac{1}{2}x^2e^x+x+2$
5. $y''+y=x^2+x$ **Ans:** $y_{PI}=x^2+x-2$
6. $y''+5y'+6=9x^4-x$ **Ans:** $y_{PI}=\frac{3}{2}x^4-5x^3+\frac{19}{2}x^2-11x+6$
7. $y''+y=2\sin x$ **Ans:** $y_{PI}=-x\cos x$
8. $y'''+2y''-y'-2y=1-4x^3$ **Ans:** $y_{PI}=2x^3-3x^2+15x-8$
9. $y''+y'-6y=52\cos 2x$ **Ans:** $y_{PI}=\sin 2x-5\cos 2x$
10. $y'''+y'+y=e^{4x}(2x+3)$ **Ans:** $y_{PI}=\frac{e^{4x}}{4761}(138x+109)$

b. Find the general solution of the following differential equations:

1. $y''+3y'+2y=6$ **Ans:** $y=ae^{-x}+be^{-2x}+3$
2. $y''-10y'+25y=30x+3$ **Ans:** $y=ae^{5x}+bxe^{5x}+\frac{6}{5}x+\frac{3}{5}$
3. $y''+y=-x-x^2$ **Ans:** $y=a\cos x+b\sin x-x^2-x+2$
4. $y''+4y=e^{-x}$ **Ans:** $y=a\cos 2x+b\sin 2x+\frac{1}{5}e^{-x}$
5. $y''+3y=-48x^2e^{3x}$ **Ans:** $y=a\cos\sqrt{3}x+b\sin\sqrt{3}x+(-4x^2+4x-\frac{4}{3})e^{3x}$
6. $y''-y'+\frac{1}{4}y=3+e^{x/2}$ **Ans:** $y=ae^{x/2}+bxe^{x/2}+12+\frac{1}{2}x^2e^{x/2}$
7. $y''+y=\sin x$ **Ans:** $y=a\cos x+b\sin x-\frac{1}{2}x\cos x$
8. $y''+y=2x\sin x$ **Ans:** $y=a\cos x+b\sin x-\frac{1}{2}x^2\cos x+\frac{1}{2}x\sin x$
9. $y''-2y'+5y=e^x\cos 2x$ **Ans:** $y=ae^x\cos 2x+be^x\sin 2x+\frac{1}{4}xe^x\sin 2x$
10. $y''-4y'+3y=e^{3x}$ **Ans:** $y=ae^{3x}+be^x+\frac{1}{2}xe^{3x}$
11. $y'''+y''-4y'-4y=3e^{-x}-4x-6$

    Ans: $y=ae^{2x}+be^{-2x}+(c-x)e^{-x}+x+\frac{1}{2}$

12. $y''+2y'+y=\sin x+3\cos 2x$

Ans: $y=ae^{-x}+bx\,e^{x}-\frac{1}{2}\cos x+\frac{12}{25}\sin 2x-\frac{9}{25}\cos 2x$

13. $y'''-6y''=3-\cos x$ **Ans**: $y=a+bx+ce^{6x}-\frac{1}{4}x^2-\frac{6}{37}\cos x+\frac{1}{37}\sin x$

14. $y'''-3y''+3y'-y=x-4e^x$

Ans: $y=ae^x+bx\,e^x+cx^2e^x-x-3-\frac{2}{3}x^3e^x$

15. $y''+y=8\sin^2 x$ **Ans**: $y=a\cos x+b\sin x+4+\frac{4}{3}\cos 2x$

c. Solve the following initial value problems:

1. $y''+4y=-2$, when $x=\frac{\pi}{8}$, $y=\frac{1}{2}$ and $y'=2$ **Ans**: $y=\sqrt{2}\sin 2x-\frac{1}{2}$

2. $5y''+y'=-6x$, when $x=0$, $y=0$ and $y'=-10$

Ans: $y=-200+200e^{-x/5}-3x^2+30x$

3. $y''+4y'+5y=35e^{-4x}$, when $x=0$, $y=-3$ and $y'=1$

Ans: $y=-10e^{-2x}\cos x+9e^{-2x}\sin x+7e^{-4x}$

4. $y''+y=\cos x-\sin 2x$, when $x=\frac{\pi}{2}$, $y=0$ and $y'=0$

Ans: $y=-\frac{1}{6}\cos x-\frac{\pi}{4}\sin x+\frac{1}{2}x\sin x+\frac{1}{3}\sin 2x$

5. $y'''-2y''+y'=2-24e^x+40e^{5x}$, when $x=0$, $y=\frac{1}{2}$, $y'=\frac{5}{2}$ and $y''=-\frac{9}{2}$

Ans: $y=11-11e^x+9x\,e^x+2x-12x^2e^x+\frac{1}{2}e^{5x}$

6. $y''-4y=2-8x$; when $x=0$, $y=0$ and $y'=5$

Ans: $y=e^{2x}-\frac{1}{2}e^{-2x}+2x-\frac{1}{2}$

7. $y''+3y'=-18x$; when $x=0$, $y=0$ and $y'=5$

Ans: $y=1+2x-3x^2-e^{-3x}$

8. $\dfrac{d^2x}{dt^2}+4\dfrac{dx}{dt}+5x=10$; when $t=0$, $x=0$ and $\dfrac{dx}{dt}=0$

Ans: $x=2(1-e^{-2t}\cos t-2e^{-2t}\sin t)$

9. $y''+4y'=2x-8$; when $x=0$, $y=1$ and $y'=0$

Ans: $4y=12\cos 2x-\sin 2x+2x-8$

3.8. The Method of Variation of Parameters

We consider again the linear second order differential equation with constant coefficients

$$y''+by'+cy=f(x) \tag{24}$$

In the previous sections, we have restricted ourselves to only few of the functions $f(x)$ that appear in the right hand side of Equation (24). These functions were $e^{\alpha x}$, $\sin\beta x$, x^2+x+1, $\cosh kx$ and the likes. What about other functions such as $\ln x$, $\frac{1}{x}$, $\sqrt{x+1}$, $etc\cdots$? In this subsection we treat such differential equations containing these functions.

We know that the solution of a differential equation is the sum of the complementary function and the particular integral. Also, we know that the complementary function is a linear combination of the two linearly independent solutions of the related homogeneous equation. Let us assume that these two linearly independent solutions y_1 and y_2 are known. Then the complementary function is

$$y_{CF}=a_1y_1+a_2y_2 \tag{25}$$

where a_1 and a_2 are the two arbitrary constants. Suppose we try a particular integral of the form

$$y=v_1y_1+v_2y_2 \tag{26}$$

where v_1 and v_2 are now two functions in x. Then, if this is true, this solution must satisfy the differential equation. Now,

$$y'=v_1y_1'+v_1'y_1+v_2y_2'+v_2'y_2=(v_1y_1'+v_2y_2')+(v_1'y_1+v_2'y_2) \tag{27}$$

Since we only said that v_1 and v_2 are functions in x, we now impose on them the condition that

$$v_1'y_1+v_2'y_2=0 \tag{28}$$

such that the form of y' is similar to that of y_{CF}' which is

$$y_{CF}'=a_1y_1'+a_2y_2'$$

Then

$$y'=v_1y_1'+v_2y_2' \tag{29}$$

and

$$y''=v_1y_1''+v_1'y_1'+v_2y_2''+v_2'y_2' \tag{30}$$

Substituting for y, y' and y'' in the differential Equation (24), we get

$$(v_1y_1''+v_1'y_1'+v_2y_2''+v_2'y_2')+b(v_1y_1'+v_2y_2')+c(v_1y_1+v_2y_2)=f(x) \tag{31}$$

Rearranging, we get

$$v_1(y_1''+by_1'+cy_1)+v_2(y_2''+by_2'+cy_2)+(v_1'y_1'+v_2'y_2')=f(x) \tag{32}$$

and since y_1 and y_2 are the two independent solutions of the related homogeneous equation, then $y_1''+by_1'+cy_1=0$ and $y_2''+by_2'+cy_2=0$ and Equation (32) becomes

$$v_1' y_1' + v_2' y_2' = f(x) \tag{33}$$

Equations (28) and (33) are two equations in two unknowns v_1' and v_2', which can be written in the form, for simplicity

$$\begin{bmatrix} y_1 & y_2 \\ y_1' & y_2' \end{bmatrix} \begin{bmatrix} v_1' \\ v_2' \end{bmatrix} = \begin{bmatrix} 0 \\ f(x) \end{bmatrix}$$

Solving for the unknowns, we get

$$v_1' = -\frac{1}{W} y_2 f(x) \text{ and } v_2' = \frac{1}{W} y_1 f(x)$$

where W is the Wronskian of y_1 and y_2 given by $W = y_1 y_2' - y_2 y_1'$.

Integrating, we obtain

$$v_1 = -\int \frac{y_2 f(x)}{W} dx + a_1 \tag{34}$$

and

$$v_2 = \int \frac{y_1 f(x)}{W} dx + a_2 \tag{35}$$

where a_1 and a_2 are the two constants of integration. If they are left then they constitute the two arbitrary constants in the general solution. The general solution of the differential equation is

$$y = a_1 y_1 + a_2 y_2 - y_1 \int \frac{y_2 f(x)}{W} dx + y_2 \int \frac{y_1 f(x)}{W} dx\,. \tag{36}$$

Notes: 1. This method assumes that we know, or can find, the two linearly independent solutions of the related homogeneous equation.

2. The method can be extended easily to higher order linear differential equations. For example, for a 3rd order equation, we will end up with three equations in three unknowns v_1', v_2' and v_3', namely

$$\left.\begin{aligned} v_1' y_1 + v_2' y_2 + v_3' y_3 &= 0 \\ v_1' y_1' + v_2' y_2' + v_3' y_3' &= 0 \\ v_1' y_1'' + v_2' y_2'' + v_3' y_3'' &= f(x) \end{aligned}\right\} \tag{37}$$

3. The method is applicable to linear differential equations with variable coefficients assuming that we know the linearly independent solutions of the related homogeneous equation. We will see this later.

To summarize, we have the following steps:

Step 1: Solve for the complementary function $y_{CF} = a_1 y_1 + a_2 y_2$.

Step 2: Compute the Wronskian $W = y_1 y_2' - y_2 y_1'$.

Step 3: Assume a particular solution of the form $y_{PI} = v_1 y_1 + v_2 y_2$.

Step 4: Compute $v_1 = -\int \frac{y_2 f(x)}{W} dx + a_1$ and $v_2 = \int \frac{y_1 f(x)}{W} dx + a_2$.

Step 5: The general solution is $y = y_{CF} + y_{PI}$.

We give here some illustrative examples.

***Example* 1**: Find the general solution of the equation $(D^2 + 1) y = \tan x$.

Solution: The complementary function is $y_{CF} = a \cos x + b \sin x$.

For the particular integral, the two linearly independent solution of the related homogeneous equation are $y_1 = \cos x$ and $y_2 = \sin x$. Then we assume a solution of the form $y = v_1 \cos x + v_2 \sin x$.

The Wronskian is

$$W = y_1 y_2' - y_2 y_1' = \cos x \cos x - \sin x (-\sin x) = 1.$$

Then the functions v_1 and v_2 are

$$v_1 = -\int \sin x \tan x \, dx = -\ln(\sec x + \tan x) + \sin x + a_1$$

and $v_2 = \int \cos x \tan x \, dx = -\cos x + a_2$.

The general solution is

$$y = \cos x [-\ln(\sec x + \tan x) + \sin x + a_1] + \sin x (-\cos x + a_2)$$

or $y = a_1 \cos x + a_2 \sin x - \cos x \ln(\sec x + \tan x)$. □

***Example* 2**: Find the general solution of $(D^2 - 3D + 2) y = \frac{1}{1 + e^{-x}}$.

Solution: The complementary function is $y_{CF} = a e^x + b e^{2x}$.

For the particular integral, the two linearly independent solution of the related homogeneous equation are $y_1 = e^x$ and $y_2 = e^{2x}$.

Then we assume a solution of the form $y = v_1 e^x + v_2 e^{2x}$.

The Wronskian is

$$W = y_1 y_2' - y_2 y_1' = e^x (2e^{2x}) - e^{2x} (e^x) = e^{3x} .$$

Then the functions v_1 and v_2 are

$$v_1 = -\int \frac{e^{2x}\, dx}{e^{3x} (1+e^{-x})} = \ln(1+e^{-x}) + a_1$$

and $$v_2 = \int \frac{e^x\, dx}{e^{3x} (1+e^{-x})} = -\tfrac{1}{2} e^{-2x} + a_2 .$$

The general solution is

$$y = a_1 e^x + a_2 e^{2x} + e^x \ln(1+e^{-x}) - \tfrac{1}{2} .$$ □

***Example* 3**: Find the general solution of the following differential equation

$$(D^2 - 6D + 9)\, y = \frac{e^{3x}}{x^2} .$$

Solution: The complementary function is

$$y_{CF} = a e^{3x} + b x e^{3x} .$$

For the particular integral, the two linearly independent solution of the related homogeneous equation are $y_1 = e^{3x}$ and $y_2 = x e^{3x}$.

Then we assume a solution of the form $y = v_1 e^{3x} + v_2 x e^{3x}$.

The Wronskian is

$$W = y_1 y_2' - y_2 y_1' = e^{3x} (3x e^{3x} + e^{3x}) - 3x e^{6x} = e^{6x} .$$

Then the functions v_1 and v_2 are

$$v_1 = -\int \frac{x e^{3x} e^{3x}}{e^{6x} x^2} dx = -\ln x + a_1$$

and $$v_2 = \int \frac{e^{3x} e^x}{e^{6x} x^2} dx = -\frac{1}{x} + a_2 .$$

The general solution is

$$y = a_1 e^{3x} + a_2 x e^{3x} - e^{3x} \ln x .$$ □

Exercise 3.6

a. Using the method of variation of parameters solve the differential equations:

1. $y''+n^2y=\sec nx$ **Ans:** $y=a\cos nx+b\sin nx-\dfrac{x\sin nx}{n}+\dfrac{\cos nx\ln(\cos nx)}{n^2}$

2. $y''-6y'+9y=\dfrac{e^{3x}}{x^2}$ **Ans:** $y=(a+bx)e^{3x}-(1+\ln x)e^{3x}$

3. $y''-y=\dfrac{2}{1+e^x}$ **Ans:** $y=ae^x+be^{-x}-1+e^x\ln\left(\dfrac{1+e^x}{e^x}\right)-e^{-x}\ln(1+e^x)$

4. $y''+y=\operatorname{cosec}x+x$ **Ans:** $y=a\cos x+b\sin x-x\cos x+\sin x\ln(\sin x)+x$

5. $y''-4y'+4y=\dfrac{e^{2x}}{x}$ **Ans:** $y=(a+bx+x\ln x-x)e^{2x}$

6. $y''+2y'+y=e^{-x}\cos x$ **Ans:** $y=(a+bx-\cos x)e^{-x}$

7. $y''-2y'+y=x^{3/2}e^x$ **Ans:** $y=(a+bx+\frac{4}{35}x^{7/2})e^x$

8. $y''+2y'+2y=\dfrac{e^{-x}}{\cos 3x}$ **Ans:** $y=e^{-x}(a\cos x+b\sin x-\frac{1}{2}\cos 2x\sec x)$

b. Verify that $y=ae^x+bx^{-1}$ is the complementary function for

$x(x+1)y''+(2-x^2)y'-(2+x)y=(x+1)^2$,

and find the general solution. **Ans:** $y=ae^x+bx^{-1}-\frac{1}{2}(x+2)$

c. Verify that $y=ae^x+bx$ is the complementary function for the differential equation: $(1-x)y''+xy'-y=2(x-1)^2e^{-x}$, and find the general solution. **Ans:** $y=ae^x+bx+e^{-x}(\frac{1}{2}-x)$

d. Verify that $y=ax+b\sin x$ is the complementary function for

$(\sin x-x\cos x)y''-(x\sin x)y'+(\sin x)y=x$,

and find the general solution. **Ans:** $y=ax+b\sin x+\cos x$

e. Verify that $y=ax+bx^2$ is the solution of the differential equation:

$x^2y''-2xy'+2y=0$, then find the general solution of:

$x^2y''-2xy'+2y=x^3$.

3.9. One Solution Is Known

Suppose we only know one solution y_1 of the related homogeneous equation, then we can still assume that the particular integral takes the form

$$y = v\,y_1 \tag{38}$$

where v is an unknown function of x. Then, if this is true, it must satisfy the differential Equation (24). Now,

$$y' = v\,y_1' + v'y_1 \text{ and } y'' = v\,y_1'' + 2v'y_1' + v''y_1 \tag{39}$$

Substituting in the differential equation (24), we get

$$v\,y_1'' + 2v'y_1' + v''y_1 + bv'y_1 + bv\,y_1' + cv\,y_1 = f(x)$$

or $$y_1v'' + (2y_1' + by_1)v' + (y_1'' + by_1' + cy_1)v = f(x) \tag{40}$$

But since y_1 is a solution of the related homogeneous equation, then

$$y_1'' + by_1' + cy_1 = 0 \tag{41}$$

and Equation (40) reduces to

$$y_1v'' + (2y_1' + by_1)v' = f(x) \tag{42}$$

Let $v' = u$, then $$y_1u' + (2y_1' + by_1)u = f(x) \tag{43}$$

Equation (43) is now a first order linear differential equation in u. Its integrating factor is

$$\mu = e^{\int [(2y_1' + by_1)/y_1]dx},$$

and the solution u is $$u = \frac{1}{\mu}\left\{ \int \frac{\mu f(x)}{y_1}dx + c_1 \right\},$$

where c_1 is the constant of integration. The solution for v is $v = \int u\,dx + c_2$.

The general solution is now $y = v\,y_1$. The function v will contain the two arbitrary constants of the solution, namely c_1 and c_2.

Notes: 1. This method will reduce the order of the equation by one, hence we can obtain the general solution of the differential equation by solving two first order linear differential equations successively.

2. The method can be applied to linear differential equations with variable coefficients if we know one solution of the related homogeneous equation.

3. The second solution of the related homogeneous equation will be obtained as a byproduct of the procedure.

***Example* 1**: Verify that sinx is a solution of the related homogeneous equation of $(D^2+1)y = \operatorname{cosec} x$ and find the general solution.

Solution: It can be easily shown that $y_1 = \sin x$ is a solution of $(D^2+1)y = 0$. Then, assume that the particular integral is of the form $y = v \sin x$. Substituting in the differential equation, we get $v'' \sin x + 2v' \cos x = \operatorname{cosec} x$.

Let $v' = u$, then $u' + 2\cot x \cdot u = \operatorname{cosec}^2 x$. This is a first order linear equation whose integrating factor is

$\mu = e^{\int 2\cot x\, dx} = e^{2\ln \sin x} = \sin^2 x$, and the solution is

$u \sin^2 x = \int \sin^2 x \operatorname{cosec}^2 x\, dx = x + a$ or $u = (x+a)\operatorname{cosec}^2 x$,

and $v = \int (x+a)\operatorname{cosec}^2 x\, dx = -(x+a)\cot x + \ln \sin x + b$.

Then the general solution is

$y = a\cos x + b\sin x - x\cos x + \sin x \ln \sin x$. □

***Example* 2**: If $y = e^{x^2}$ is a solution of the related homogeneous equation of $xy'' - y' - 4x^2 y = -4x^5$, find its general solution.

Solution: Assuming a general solution of the form $y = v\, e^{x^2}$, then

$y' = 2xv\, e^{x^2} + e^{x^2} v' = e^{x^2}(2xv + v')$ and

$$y'' = e^{x^2}(2xv' + 2v + v'') + 2x\, e^{x^2}(2xv + v')$$
$$= e^{x^2}[v'' + 4xv' + 2(2x^2+1)v].$$

Substituting in the differential equation, we obtain

$$xe^{x^2}[v'' + 4xv' + 2(2x^2+1)v] - e^{x^2}[v' + 2xv] - 4x^2 e^{x^2} v = -4x^5$$

or $v'' - \dfrac{4x^2-1}{x} v' = -4x^4 e^{-x^2}$.

Letting $u = v'$, we get $u' - \dfrac{4x^2-1}{x} u = -4x^4 e^{-x^2}$

This is a linear first order equation whose integrating factor is

$\mu = e^{\int \frac{4x^2-1}{x} dx} = \dfrac{e^{2x^2}}{x}$, and its solution is

$$u = 2x e^{-x^2}(1-x^2)+c_1 x e^{-2x^2}.$$

The obtain *v*, we integrate *u* to get

$$v = x^2 e^{-x^2} - \tfrac{1}{4}c_1 e^{-2x^2} + c_2$$

The general solution of the differential equation is

$$y = (x^2 e^{-x^2} - \tfrac{1}{4}c_1 e^{-2x^2} + c_2)e^{x^2} = x^2 - \tfrac{1}{4}c_1 e^{-x^2} + c_2 e^{x^2}. \square$$

3.10. Relation Between the Two Solutions

Consider the second order linear homogeneous differential equation with constant coefficients

$$y''+by'+cy=0 \tag{44}$$

We know that there is two independent solutions y_1 and y_2, i.e. the Wronskian of the two solutions should not vanish

$$W=\begin{vmatrix} y_1 & y_2 \\ y_1' & y_2' \end{vmatrix} \neq 0 \tag{45}$$

Now, $W=y_1y_2'-y_2y_1'=y_1^2\dfrac{d}{dx}\left(\dfrac{y_2}{y_1}\right)$ then

$$\frac{d}{dx}\left(\frac{y_2}{y_1}\right)=\frac{W}{y_1^2} \tag{46}$$

Integrating, we obtain

$$y_2=y_1\int\frac{W}{y_1^2}dx \tag{47}$$

Now, suppose that only one solution y_1 is known, can we determine the second solution y_2 from relation (47)? We have to determine the Wronskian first. To find the Wronskian W we say that since y_1 and y_2 are two solutions of Equation (44), they must satisfy the differential equation, i.e.,

$$y_1''+by_1'+cy_1=0 \text{ and } y_2''+by_2'+cy_2=0 \tag{48}$$

Multiplying the first equation by y_2 and the second by y_1 and subtracting, we get

$$y_1y_2''-y_2y_1''+b(y_1y_2'-y_2y_1')=0$$

or

$$\frac{d}{dx}(y_1y_2'-y_2y_1')+b\,(y_1y_2'-y_2y_1')=0$$

or

$$\frac{d}{dx}W+bW=0 \tag{49}$$

Equation (49) is a first order differential equation in W, the solution is

$$W=e^{-\int b\,dx}=e^{-bx} \tag{50}$$

where the constant of integration is omitted. Then the second solution is

$$y_2=y_1\int\frac{e^{-bx}}{y_1^2}dx \tag{51}$$

Note: This procedure may as well be applied to linear homogeneous equation with variable coefficients. If y_1 is a solution of

$$y''+P(x)y'+Q(x)y=0$$

then the second solution is $y_2 = y_1 \int \frac{e^{-\int P(x)dx}}{y_1^2} dx$ (52)

***Example* 1**: if $y=e^{-x}$ is a solution of the equation $y''+2y'+y=0$, find the general solution.

Solution: The second solution is given by $y_2 = e^{-x} \int e^{2x} e^{-2x} dx = x e^{-x}$.

The general solution is $y=(a+bx)e^{-x}$. □

***Example* 2**: If $y=x-1$ is a solution of the differential equation $x(x-1)y''+(x-1)y'-y=0$, find the general solution.

Solution: The second solution is given by

$$y_2 = (x-1)\int \frac{e^{-\int \frac{1}{x}dx}}{(x-1)^2} dx = (x-1)\int \frac{dx}{x(x-1)^2} = (x-1)\ln\left(\frac{x}{x-1}\right) - 1$$

(we have used partial fractions)

The general solution is $y = a(x-1) + b\left\{\ln\left(\frac{x}{x-1}\right) - 1\right\}$. □

***Example* 3**: Show that $y=e^{\sin^{-1}x}$ satisfies the differential equation: $(1-x^2)y''-xy'-y=0$, then find its general solution.

Solution: If $y=e^{\sin^{-1}x}$, then $y' = \frac{e^{\sin^{-1}x}}{\sqrt{1-x^2}}$ or $\sqrt{1-x^2}\,y' = y$. Then differentiating, we get $\sqrt{1-x^2}\,y'' - \frac{xy'}{\sqrt{1-x^2}} - y' = 0$ or

$$\sqrt{1-x^2}\,y'' - \frac{xy'}{\sqrt{1-x^2}} - \frac{y}{\sqrt{1-x^2}} = 0.$$

Multiplying by $\sqrt{1-x^2}$ we obtain $(1-x^2)y''-xy'-y=0$.

Then $y=e^{\sin^{-1}x}$ satisfies the differential equation and is the first solution. To obtain the second solution, we have

$$y_2 = y_1 \int \frac{e^{-\int P(x)dx}}{y_1^2} dx = e^{\sin^{-1} x} \int \frac{e^{\int \frac{x}{1-x^2} dx}}{e^{2\sin^{-1} x}} dx$$

$$= e^{\sin^{-1} x} \int \frac{e^{\ln(1/\sqrt{1-x^2})}}{e^{2\sin^{-1} x}} dx = e^{\sin^{-1} x} \int \frac{e^{-2\sin^{-1} x}}{\sqrt{1-x^2}} dx$$

$$= e^{\sin^{-1} x} \int e^{-2\sin^{-1} x}\, d \sin^{-1} x = -\frac{1}{2} e^{-\sin^{-1} x}$$

The general solution is $y = A\, e^{\sin^{-1} x} + B\, e^{-\sin^{-1} x}$. □

3.11. Euler Homogeneous Equation

This particular equation is not with constant coefficients but, with a special substitution for the independent variable, it can be transformed to a linear equation with constant coefficients. An nth order Euler homogeneous equations takes the form

$$a_0 x^n \frac{d^n y}{dx^n} + a_1 x^{n-1} \frac{d^{n-1} y}{dx^{n-1}} + \cdots + a_n y = f(x)$$

Without loss of generality and out of simplification we will consider here the second order Euler equation (also called Cauchy equation) in the form

$$x^2 y'' + bx\, y' + cy = f(x) \tag{53}$$

Again, we should not be confused with the term "*homogeneous*" here, it is not what it meant to be in the past. It means that the first derivative of y is multiplied by x, the second by x^2, and so on for higher order equations.

Using the substitution $x = e^t$, we get

$$\frac{dy}{dx} = \frac{dy}{dt} \cdot \frac{dt}{dx} = e^t \frac{dy}{dt} \quad \text{or simply} \quad x\frac{dy}{dx} = \frac{dy}{dt} \tag{54}$$

Differentiating again, we get

$$x\frac{d^2 y}{dx^2} + \frac{dy}{dx} = \frac{d^2 y}{dt^2} \cdot \frac{dt}{dx} = \frac{1}{x}\frac{d^2 y}{dt^2}$$

or

$$x^2 \frac{d^2 y}{dx^2} = \frac{d^2 y}{dt^2} - \frac{dy}{dt} \tag{55}$$

In operator form where $D \equiv \frac{d}{dx}$ and $\Theta \equiv \frac{d}{dt}$, we have

$$xDy = \Theta y \quad \text{and} \quad x^2 D^2 y = \Theta(\Theta - 1)y \tag{56}$$

and Equation (53) becomes

$$[\Theta(\Theta-1) + b\Theta + c]y = f(e^t)$$

or

$$[\Theta^2 + (b-1)\Theta + c]y = f(e^t) \tag{57}$$

Equation (57) is a linear differential equation with constant coefficients where y is the dependent variable and t is now the independent variable which can be solved using the techniques given in the previous subsections.

For n^{th} order Euler equation, we have

$$x^n D^n y = [\Theta(\Theta-1)(\Theta-2)\cdots(\Theta-n+1)]\,y, \; n = 1, 2, 3, \cdots \tag{58}$$

***Example* 1**: Solve the equation $(x^2D^2 + xD - 1)y = 0$.

Solution: Let $x = e^t$, the equation is transformed to

$$[\Theta(\Theta-1)+\Theta-1]y = 0 \quad \text{or} \quad (\Theta^2-1)y = 0.$$

Then $y = ae^t + be^{-t}$ or $y = ax + bx^{-1}$. □

***Example* 2**: Solve the equation $(x^2D^2 - xD + 1)y = 2\ln x$.

Solution: Let $x = e^t$, the equation is transformed to

$$[\Theta(\Theta-1)-\Theta+1]y = 2t \quad \text{or} \quad (\Theta^2-2\Theta+1)y = 2t.$$

The complementary function of this equation is $y_{CF} = e^t(a+bt)$.

The particular integral can be found to be $y_{PI} = 2t + 4$.

From which $y = e^t(a+bt) + 2t + 4$

or $y = ax + bx\ln x + 2\ln x + 4$. □

***Example* 3**: Solve the equation $(x^2D^2 + 4xD + 2)y = \sin(\ln x)$.

Solution: Let $x = e^t$, the equation is transformed to

$$[\Theta(\Theta-1)+4\Theta+2]y = \sin t \quad \text{or} \quad (\Theta^2+3\Theta+2)y = \sin t.$$

The solution can be found to be

$$y = ae^{-2t} + be^{-t} + \tfrac{1}{10}(\sin t - 3\cos t),$$

or $y = ax^{-2} + bx^{-1} + \tfrac{1}{10}[\sin(\ln x) - 3\cos(\ln x)]$. □

***Example* 4**: Solve the equation $(x^2D^2 - xD + 4)y = \cos(\ln x) + x\sin(\ln x)$.

Solution: Let $x = e^t$, the equation is transformed to

$$[\Theta(\Theta-1)-\Theta+4]y = \cos t + e^t\sin t$$

or $(\Theta^2 - 2\Theta + 4)y = \cos t + e^t \sin t$.

The solution can be found to be

$y = e^t[a\cos(\sqrt{3}t) + b\sin(\sqrt{3}t)] + \tfrac{1}{13}[3\cos t - 2\sin t] + \tfrac{1}{2}e^t\sin t$, or

$$y = x[a\cos(\sqrt{3}\ln x) + b\sin(\sqrt{3}\ln x)]$$
$$+\tfrac{1}{13}[3\cos(\ln x) - 2\sin(\ln x)] + \tfrac{1}{2}x\sin(\ln x).$$ □

Exercise 3.7

1. $x^2y''+xy'-y=4$ **Ans:** $y=ax^{-1}+bx-4$

2. $x^2y''-2xy'+2y=x^4$ **Ans:** $y=ax+bx^2+\frac{1}{6}x^4$

3. $x^2y''-2xy'+2y=\dfrac{1}{x^2}$ **Ans:** $y=ax+bx^2+\dfrac{1}{12x^2}$

4. $x^2y''-4xy'+6y=x^{-4}$ **Ans:** $y=ax^3+bx^2+\dfrac{1}{42x^4}$

5. $xy''-y'=2x^2e^x$ **Ans:** $y=a+bx+(2x-2)e^x$

6. $x^2y''+5xy'+3y=\dfrac{(1+x)^2\ln x}{x^2}$

Ans: $y=ax^{-3}+bx^{-1}+\frac{1}{9}(3\ln x-4)+\dfrac{1}{2x}(\ln^2 x-\ln x)-\dfrac{1}{x^2}\ln x$

7. $x^3y'''+3x^2y''-2xy'+2y=\ln x$

Ans: $y=x(a+b\ln x)+cx^{-2}+\frac{1}{2}\ln x+\frac{3}{4}$

8. $(x+2)^2y''+(x+2)y'-y=x$

Ans: $y=a(x+2)+\dfrac{b}{x+2}+\frac{1}{2}(x+2)\ln(x+2)+2$

Hint: put $x+2=e^t$. This type is sometimes called Legendre linear equation.

9. $(x+1)^2y''+(x+1)y'-y=\ln(1+x)^2+x-1$

Ans: $y=\dfrac{a}{x+1}+\frac{1}{2}(x+1)\ln(x+1)-\ln(x+1)^2+2$

10. $x^2y''-2y=x$, when $x=1$, $y=0$ and $y'=0$

Ans: $y=\frac{1}{3}x^2+\dfrac{1}{6x}-\frac{1}{2}x$

11. $x^3y'''+3x^2y''+xy'+8y=65\cos(\ln x)$

Ans: $y=\dfrac{a}{x^2}+x[b\cos(\sqrt{3}\ln x)+c\sin(\sqrt{3}\ln x)]+8\cos(\ln x)-\sin(\ln x)$

12. Using the substitution $y=t^2$, solve the differential equation:

$2x^2y\,y''+4y^2-x^2y'^2-2x\,y\,y'=0$ **Ans:** $y=(a+b\ln x)^2x^2$

3.12. System of Simultaneous Linear Equations

In this section, we restrict ourselves to linear differential systems with constant coefficients. Let us consider the following system

$$\left.\begin{aligned} y''-y+5x' &= t \\ 2y'-x''+4x &= 2 \end{aligned}\right\} \tag{59}$$

Here ′ and ″ represent differentiation with respect to the independent variable t, while x and y are the two dependent variables. In operator form, where $D \equiv \frac{d}{dt}$, we may write

$$\left.\begin{aligned} (D^2-1)y+5Dx &= t \\ 2Dy-(D^2-4)x &= 2 \end{aligned}\right\} \tag{60}$$

To solve this system, we use the operational method. The idea is to eliminate one variable (x or y) from one of the two equations. We may, for instance operate on the first equation with $2D$ and on the second equation with (D^2-1) and then subtract one from the other to get

$$[10D^2+(D^2-1)(D^2-4)]x = 2D(t)-(D^2-1)(2)$$

or

$$(D^4+5D^2+4)x = 4 \tag{61}$$

In a similar manner we could have eliminated x. The resultant equation for y is

$$[(D^2-1)(D^2-4)+10D^2]y = (D^2-4)(t)+5D(2)$$

or

$$(D^4+5D^2+4)y = -4t \tag{62}$$

Equations (61) and (62) represent two differential equations in x and the other in y, they can be solved independently. Their solutions are

$$x = a_1\cos t + a_2\sin t + a_3\cos 2t + a_4\sin 2t + 1 \tag{63}$$

and

$$y = b_1\cos t + b_2\sin t + b_3\cos 2t + b_4\sin 2t - t \tag{64}$$

Equations (63) and (64) represent the solution of the system. However, we notice the presence of eight constants. How come? We would expect to have only four constants. Then there must be a linear relation between these eight constants to reduce them to only four. To find this relation, x and y as given by Equations (63) and (64) must satisfy the differential system (60). Then substituting for x and y in the first Equation of (60), we obtain the identity

$$t-2b_1\cos t-2b_2\sin t-5b_3\cos 2t-5b_4\sin 2t$$

$$-5a_1\sin t+5a_2\cos t-10a_3\sin 2t+10a_4\cos 2t \equiv -t$$

Then, we must have

$$\left.\begin{aligned} -2b_1+5a_2&=0\\ -2b_2-5a_1&=0\\ -5b_3+10a_4&=0\\ -5b_4-10a_3&=0 \end{aligned}\right\} \qquad (65)$$

Substituting, in Equation (64), for the *b*'s in terms of the *a*'s using equation (65), we obtain the solution of the system as

$$x = a_1\cos t + a_2\sin t + a_3\cos 2t + a_4\sin 2t + 1 \qquad (66)$$

and
$$y = \tfrac{5}{2}a_2\cos t - \tfrac{5}{2}a_1\sin t + 2a_4\cos 2t - 2a_3\sin 2t - t \qquad (67)$$

A simpler technique for solving the system of equations (60) is given as follows:

1. Obtain Equation (61) and solve it for x to give Equation (63)
2. Rewrite the system (60) as

$$(2D^2-2)y + 10Dx = 2t \qquad \text{(multiply by 2)}$$
$$2D^2y - (D^3-4D)x = 0 \qquad \text{(operate with } D\text{)}$$

Subtracting, we get $2y - D^3x - 6Dx = -2t$,

or
$$y = \tfrac{1}{2}D^3x + 3Dx - t \qquad (68)$$

3. Substituting for x from Equation (63) in Equation (68) we obtain the solution for y as in Equation (67).

A method that proved to be very effective in solving systems of linear differential equations with constant coefficients is the Laplace transform method; but this is another story.

***Example* 1**: Solve the system of equations: $\begin{aligned} x' &= 4x-2y\\ y' &= x+y \end{aligned}$.

Solution: In operator form, we have $\begin{aligned} (D-4)x+2y&=0 \qquad (i)\\ x-(D-1)y&=0 \qquad (ii) \end{aligned}$

Operating on equation (*ii*) with $(D-4)$ and subtracting, we obtain

$2y + (D-4)(D-1)y = 0$ or $(D^2-5D+6)y = 0$,

hence $y = a_1e^{3t} + a_2e^{2t}$

Substituting for y in Equation (*ii*), we get

$x - (D-1)(a_1e^{3t} + a_2e^{2t}) = 0$ or $x = 2a_1e^{3t} + a_2e^{2t}$.

The solution is $y = a_1e^{3t} + a_2e^{2t}$ and $x = 2a_1e^{3t} + a_2e^{2t}$. □

***Example* 2**: Solve the system of equations: $\begin{array}{l} x' - y = t \\ y' + x = 1 \end{array}$.

Solution: In operator form, we have $\begin{array}{ll} Dx - y = t & (i) \\ x + Dy = 1 & (ii) \end{array}$

Differentiating Equation equation (*i*), we get $D^2x - Dy = 1$.

Eliminating *y* beyween this equation and (*ii*), we obtain

$D^2x + x = 2$ or $(D^2 + 1)x = 2$, whose solution is

$x = a_1 \cos t + a_2 \sin t + 2$

Then from the solution *x* and equation (*i*), we obtain

$y = Dx - t = -a_1 \sin t + a_2 \cos t - t$

Then, the general solution is

$y = -a_1 \sin t + a_2 \cos t - t$ and $x = a_1 \cos t + a_2 \sin t + 2$ □

***Example* 3**: Solve the initial value problem:

$x'' + x - y' = 0$ and $4x' + 2x - y' - 2y = 0$,

when $t = 0, x = 0, y = 1$ and $x' = 2$.

Solution: In operator form, we have

$(D^2 + 1)x - Dy = 0$ (*i*)

$(4D + 2)x - (D + 2)y = 0$ (*ii*)

Operating on equation (*i*) with $(D + 2)$ and equation (*ii*) with *D* and subtracting, we obtain

$[(D + 2)(D^2 + 1) - D(4D + 2)]x = 0$ or $(D^3 - 2D^2 - D + 2)x = 0$

Hence $x = ae^t + be^{-t} + ce^{2t}$.

Subtracting (*ii*) from (*i*), we get

$[(D^2 + 1) - (4D + 2)]x + 2y = 0$

Then, *y* can be found from

$y = -\frac{1}{2}(D^2 - 4D - 1)(ae^t + be^{-t} + ce^{2t}) = \frac{1}{2}(4ae^t - 4be^{-t} + 5ce^{2t})$

and the general solution is

$x = ae^t + be^{-t} + ce^{2t}$ and $y = \frac{1}{2}(4ae^t - 4be^{-t} + 5ce^{2t})$.

To satisfy the initial condition, we have $x' = ae^t - be^{-t} + 2ce^{2t}$.

At $t = 0$, we have

$0 = a + b + c,\ 1 = 2a - 2b + \frac{5}{2}c$ and $2 = a - b + 2c$.

Solving these three equations for the three unknowns a, b and c, we get $a = -2,\ b = 0$ and $c = 2$.

The solution of the initial value problem is

$x = -2e^t + 2e^{2t}$ and $y = -4e^t + 5e^{2t}$.

If x and y are the cartesian coordinates of a point on a curve, then these two equations represent the parametric equations of this curve, where t is the parameter. Eliminating t from the two equations we obtain the cartesian equation of the curve given by

$(5x - 2y)^2 = 4(y - 2x)$ (a parabola). □

***Example* 4**: Solve the initial value problem

$(D - 5)x - 8y = 1$ (*i*)

$6x + (D + 9)y = t$ (*ii*)

when $t = 0, x = 4$ and $y = -3$.

Solution: Operating on (*ii*) with $(D - 5)$ and multiplying (*i*) by 6 and subtraction, we obtain

$(D - 5)(D + 9)y + 48y = (D - 5)(t) - 6$ or

$(D^2 + 4D + 3)y = -5(1 + t)$

The solution of this equation can be found to be

$y = a_1 e^{-3t} + a_2 e^{-t} - \frac{5}{3}t + \frac{5}{9}$.

Substituting for y in equation (*ii*), we finally get

$x = -a_1 e^{-3t} - \frac{4}{3}a_2 e^{-t} - \frac{5}{9} + \frac{8}{3}t$.

To satisfy the initial conditions, we have at $t = 0$

$4 = -a_1 - \frac{4}{3}a_2 - \frac{5}{9}$ and $-3 = a_1 + a_2 + \frac{5}{9}$

from which $a_1 = -\frac{5}{9}$ and $a_2 = -3$, and the solution of the initial value problem is

$x = \frac{5}{9}e^{-3t} + 4e^{-t} - \frac{5}{9} + \frac{8}{3}t$ and $y = -\frac{5}{9}e^{-3t} - 3e^{-t} + \frac{5}{9} - \frac{5}{3}t$. □

Exercise 3.8

a. Solve the following systems of differential equations:

1. $x' = y,\ y' = x$ **Ans**: $x = a_1 e^t + a_2 e^{-t}$ and $y = a_1 e^t - a_2 e^{-t}$

2. $x' + y' = \cos t,\ x' - y' = \sin t$

Ans: $x = \frac{1}{2}(\sin t - \cos t) + a_1,\ y = \frac{1}{2}(\sin t + \cos t) + a_2$

3. $x'' = y + 1,\ y'' = x + t$ **Ans**: $x = a_1 e^t + a_2 e^{-t} + a_3 \cos t + a_4 \sin t - t$
$y = a_1 e^t + a_2 e^{-t} - a_3 \cos t - a_4 \sin t - 1$

4. $x' + ny = 0,\ nx - y' = 0$ **Ans**: $x = a \cos nt + b \sin nt,\ y = a \sin nt - b \cos nt$

5. $2x' + 36x + y' + 3y = -9e^{2t},\ 5x' + 22x + y' + 2y = 18e^{-t}$

Ans: $x = a_1 e^{-t} + a_2 e^{2t} + 4t(e^{2t} - e^{-t})$
$y = (17a_1 - 30)e^{-t} + (3 - 8a_2)e^{2t} - 4t(8e^{2t} - 17e^{-t})$

6. $v' - 2v + 2w' = 2 - 4e^{2t},\ 2v' - 3v + 3w' - w = 0$

Ans: $v = a_1 e^t + a_2 e^{-2t} + 5e^{2t} - 1,\ w = \frac{1}{2}a_1 e^t - a_2 e^{-2t} - e^{2t} + 3$

7. $(D^2 - 3D)y - (D - 2)x = 14t + 7,\ (D - 3)y + Dx = 1$

Ans: $y = 2 + a_1 e^t + a_2 e^{3t} + a_3 e^{-2t},\ x = 7 + 7t + 2a_1 e^t - \frac{5}{2}a_3 e^{-2t}$

8. $(2D^2 + 5)x + (D^2 + 3)y = -8\sin 3t,\ (D^2 + 7)x + (D^2 + 5)y = 8\sin 3t$

Ans: $x = 2a_1 \cos t + 2a_2 \sin t + a_3 \cos 2t + a_4 \sin 2t + 2\sin 3t$
$y = -3(a_1 \cos t + a_2 \sin t + a_3 \cos 2t + a_4 \sin 2t + \sin 3t)$

b. Solve the following initial value problems:

1. $(D - 2)x - 3y = 2e^{2t},\ -x + (D - 4)y = 3e^{2t}$

when $t = 0, x = -\frac{2}{3}$ and $y = \frac{1}{3}$ **Ans**: $x = e^{5t} - \frac{5}{3}e^{2t},\ y = e^{5t} - \frac{2}{3}e^{2t}$

2. $(D^2 + 1)y + 4(D - 1)x = 4e^t,\ (D - 1)y + (D + 9)x = 0$

when $t = 0, y = 5, y' = 0$ and $x = \frac{1}{2}$.

Ans: $x = \frac{1}{2}e^{-t} + e^{-2t}\sin t,\ y = 2e^t + 2e^{-t} + e^{-2t}(\cos t + 2\sin t)$

3. $x'' + 2x - y' = 2t + 5,\ x' - x + y' + y = -2t - 1$

when $t = 0, x = 3, x' = 0$ and $y = -3$.

Chapter Four

Linear Equations with Variable Coefficients

$$\frac{d^2y}{dx^2}+P(x)\frac{dy}{dx}+Q(x)y=f(x)$$

Chapter 4.

Linear Equations with Variable Coefficients

4.1. Introduction

These equations take the general form

$$a_0(x)\frac{d^n y}{dx^n}+a_1(x)\frac{d^{n-1}y}{dx^{n-1}}+\cdots+a_n(x)y=f(x) \qquad (1)$$

where the coefficients are functions of the independent variable x. Methods for solving these types of equations are some how limited. We consider here second order equations of the form

$$\frac{d^2y}{dx^2}+P(x)\frac{dy}{dx}+Q(x)y=f(x) \qquad (2)$$

In the following subsections, we present some methods for the solution of Equation (2). Each of these methods treats a special case under consideration.

4.1. Changing the Dependent Variable

Let us assume that the solution of Equation (2) is the product of two functions u and v of x, i.e.,

$$y=uv \qquad (3)$$

If we can obtain u and v by some technique, the solution y will be determined by Equation (3). Now,

$$\frac{dy}{dx}=u\frac{dv}{dx}+v\frac{du}{dx} \quad \text{and} \quad \frac{d^2y}{dx^2}=u\frac{d^2v}{dx^2}+2\frac{du}{dx}\cdot\frac{dv}{dx}+v\frac{d^2u}{dx^2},$$

or

$$y'=uv'+vu' \quad \text{and} \quad y''=uv''+2u'v'+vu''.$$

Substituting in the differential equation (2) and rearranging, we obtain

$$v''+\frac{1}{u}(2u'+Pu)v'+\frac{1}{u}(u''+Pu'+Qu)v=\frac{f(x)}{u} \qquad (4)$$

<u>Case I</u>: One Solution of the Related Homogeneous Equation Is Known

If some how we know a function u such that

$$u''+Pu'+Qu=0 \qquad (5)$$

i.e., u is a solution of the related homogeneous equation; then Equation (4) reduces to

$$v'' + \frac{1}{u}(2u' + Pu)v' = \frac{f(x)}{u} \tag{6}$$

Letting $v' = p$ in Equation (6), we get

$$p' + \frac{1}{u}(2u' + Pu)p = \frac{f(x)}{u} \tag{7}$$

Equation (7) is a first order linear equation where p is the dependent variable and x is the independent variable. Solving this equation for p, we obtain

$$p = \frac{1}{\mu}\left\{ \int \mu \frac{f(x)}{u}\, dx + c_1 \right\} \tag{8}$$

where μ is the integrating factor for the first order equation and is given by

$$\mu = e^{\int \frac{1}{u}(2u' + Pu)dx} \tag{9}$$

and the solution v of Equation (6) will be

$$v = \int p\, dx + c_2 . \tag{10}$$

The procedure can be summarized as follows:

Step 1: Let $y = uv$, where u is a solution of $u'' + Pu' + Qu = 0$.

Step 2: Substitute for u in $v'' + \frac{1}{u}(2u' + Pu)v' = \frac{f(x)}{u}$.

Step 3: Let $p = v'$ and solve the equation for p: $p' + \frac{1}{u}(2u' + Pu)p = \frac{f(x)}{u}$

we get $p = \frac{1}{\mu}\left\{ \int \mu \frac{f(x)}{u}\, dx + c_1 \right\}$, where $\mu = e^{\int \frac{1}{u}(2u' + Pu)dx}$.

Step 4: The solution for v is $v = \int p\, dx + c_2$

Step 5: The general solution is $y = uv$.

***Example* 1**: Show that e^x is a solution of the related homogeneous equation of

$y'' - \frac{2x+1}{x} y' + \frac{x+1}{x} y = \frac{(x^2 + x - 1)e^{2x}}{x}$, then find the general solution.

Solution: Substituting for $u = e^x$ in the related homogeneous equation, we get

$$y'' - \frac{2x+1}{x} y' + \frac{x+1}{x} y = e^x - \frac{2x+1}{x} e^x + \frac{x+1}{x} e^x = 0 .$$

Then $u = e^x$ is a solution for the related homogeneous equation.

Let $y = uv = v\,e^x$, the differential equation reduces to

$$v'' - \frac{1}{x}v' = \left(x + 1 - \frac{1}{x}\right)e^x .$$

Let $p = v'$, we get $p' - \frac{1}{x}p = \left(x + 1 - \frac{1}{x}\right)e^x$.

This is a linear first order equation whose integrating factor is

$$\mu = e^{\int -\frac{1}{x}dx} = \frac{1}{x}.$$

The solution for p is $p = x\int\left(1 + \frac{1}{x} - \frac{1}{x^2}\right)e^x\,dx = x\,e^x + e^x + c_1 x$.

The solution for v is $v = \int(x\,e^x + e^x + c_1 x)dx = x\,e^x + \frac{c_1}{2}x^2 + c_2$.

Finally, the general solution is $y = x\,e^{2x} + Ax^2 e^x + B\,e^x$. □

Example 2: Show that sin*x* is a solution of the related homogeneous equation of $y'' - (2\tan x)y' + 3y = 2\sec x$, then find the general solution.

Solution: Substituting for $u = \sin x$ in the related homogeneous equation, we get $u'' - (2\tan x)u' + 3u = -\sin x - 2\tan x\,\cos x + 3\sin x = 0$.

Then $u = \sin x$ is a solution for the related homogeneous equation.

Let $y = uv = v\,\sin x$ the differential equation reduces to

$$v'' + 2(\cot x - \tan x)v' = 4\operatorname{cosec} 2x .$$

Let $p = v'$, we get $p' + 2(\cot x - \tan x)p = 4\operatorname{cosec} 2x$.

This is a linear first order equation. The integrating factor is

$$\mu = e^{\int 2(\cot x - \tan x)dx} = \tfrac{1}{4}\sin^2 2x .$$

The solution for p is $p = -2\operatorname{cosec} 2x\,\cot 2x + c_1\operatorname{cosec}^2 2x$.

The solution for v is $v = \operatorname{cosec} 2x + c_1\cot 2x + c_2$.

Finally, the general solution is

$$y = \tfrac{1}{2}\sec x + c_1(\cos x - \tfrac{1}{2}\sec x) + c_2\sin x .$$ □

***Example* 3**: Show that e^{x^2} is a solution of the related homogeneous equation of $xy'' - y' - 4x^3 y = -4x^5$, then find the general solution.

Solution: Let $u = e^{x^2}$, then $u' = 2x\, e^{x^2}$.

Substituting in the related homogeneous equation, we get

$x(4x^2 e^{x^2} + 2e^{x^2}) - 2xe^{x^2} - 4x^3 e^{x^2} = 0$. Then $u = e^{x^2}$ is a solution for the related homogeneous equation.

Let $y = uv = v\, e^{x^2}$, then $y' = e^{x^2} v' + 2x\, e^{x^2} v$ and

$y'' = e^{x^2} [v'' + 4xv' + 2(2x^2 + 1)v]$.

Substituting in the differential equation and rearranging, we get

$$v'' + \frac{1}{x}(4x^2 - 1)v' = -4x^4 e^{-x^2}.$$

Let $p = v'$, we obtain $p' + \frac{1}{x}(4x^2 - 1)\, p = -4x^4 e^{-x^2}$.

This is a linear first order equation, its solution is found to be

$$p = 2x\,(1 - x^2) e^{-x^2} + c_1 x\, e^{-2x^2}.$$

Integrating this last expression, we obtain v as

$$v = c_2 - \frac{c_1}{4} e^{-2x^2} + x^2 e^{-x^2}.$$

Then the general solution is given by

$$y = e^{x^2} \left(c_2 - \frac{c_1}{4} e^{-2x^2} + x^2 e^{-x^2} \right) \quad \text{or}$$

$$y = A\, e^{x^2} + B\, e^{-x^2} + x^2.$$ □

The previous method applies only if we know a solution of the related homogeneous equation. Now, if no solution is known, can we find a way? Let us see. Consider the homogeneous equation

$$y'' + P(x)y' + Q(x)y = 0 \tag{11}$$

The following cases are just stated and the proof is almost obvious:

1. If $P + xQ = 0$, then $u = x$ is a solution.
2. If $2 + 2xP + x^2 Q = 0$, then $u = x^2$ is a solution.
3. If $m(m-1) + mxP + x^2 Q = 0$, then $u = x^m$ is a solution.

4. If $1+P+Q=0$, then $u=e^{x}$ is a solution.

5. If $1-P+Q=0$, then $u=e^{-x}$ is a solution.

6. If $m^2+mP+Q=0$, then $u=e^{mx}$ is a solution.

These special cases may sometimes help in identifying one solution of the related homogeneous equation.

***Example* 4**: Solve the differential equation

$$x^2(x+1)y''-x(2+4x+x^2)y'+(2+4x+x^2)y=-x^4-2x^3.$$

Solution: Here $P+xQ=-\dfrac{x(2+4x+x^2)}{x^2(x+1)}+x\dfrac{2+4x+x^2}{x^2(x+1)}=0$, then $u=x$ is a solution of the related homogeneous equation.

Let $y=xv$, differentiating twice, substituting in the differential equation and rearranging, we obtain $v''-\dfrac{x+2}{x+1}v'=-\dfrac{x+2}{x+1}$.

Let $p=v'$, then $p'-\dfrac{x+2}{x+1}p=-\dfrac{x+2}{x+1}$.

This is a linear first order equation whose integrating factor is

$$\mu=e^{-\int\left(1+\frac{1}{x+1}\right)dx}=\frac{e^{-x}}{x+1}.$$

Hence, $\dfrac{e^{-x}}{x+1}p=-\displaystyle\int\frac{e^{-x}(x+2)dx}{(x+1)^2}=\frac{e^{-x}}{x+1}+c_1$.

Then the solution p is $p=1+c_1(x+1)e^{x}$.

and v is given by $v=x+c_1xe^{x}+c_2$.

Finally, the general solution of the differential equation is

$$y=c_1x^2e^{x}+c_2x+x^2.$$ □

***Example* 5**: Solve the differential equation

$$(x+2)y''-(2x+5)y'+2y=(x+1)e^{x}.$$

Solution: $P=-\dfrac{2x+5}{(x+2)}$ and $Q=\dfrac{2}{(x+2)}$, here

$4+2P+Q=4-2\dfrac{2x+5}{x+2}+\dfrac{2}{x+2}=0$, then $u=e^{2x}$ is a solution of the related homogeneous equation.

Let $y = e^{2x} v$, differentiating twice, substituting in the differential equation and rearranging, we obtain $v'' + \frac{2x+3}{x+2} v' = \frac{x+1}{x+2} e^{-x}$.

Let $p = v'$, then $p' + \frac{2x+3}{x+2} p = \frac{x+1}{x+2} e^{-x}$.

This is a linear first order equation whose integrating factor is

$$\mu = e^{-\int \left(-2+\frac{1}{x+2}\right)dx} = \frac{e^{2x}}{x+2}.$$

Hence, $\frac{e^{2x}}{x+2} p = -\int \frac{e^x (x+1)dx}{(x+2)^2} = \frac{e^x}{x+2} + c_1$.

Then the solution p is $p = e^{-x} + c_1(x+2)e^{-2x}$.

and v is given by $v = -e^{-x} + \frac{c_1}{4}(2x+5) + c_2$.

Finally, the general solution of the differential equation is

$y = a(2x+5)e^{2x} + b e^{2x} - e^x$. □

***Example* 6**: Solve the equation $y'' - (\cot x)y' - (1-\cot x)y = e^x \sin x$.

Solution: Here $1 + P + Q = 1 - \cot x - 1 + \cot x = 0$, then $u - e^x$ is a solution of the related homogeneous equation.

Let $y = e^x v$, then $y' = e^x (v' + v)$, $y'' = e^x (v'' + 2v' + v)$.

Substituting in the differential equation and rearranging, we obtain

$v'' + (2 - \cot x)v' = \sin x$.

Let $p = v'$, then $p' + (2 - \cot x)p = \sin x$.

This is a linear first order equation whose integrating factor is

$\mu = e^{\int (2-\cot x)dx} = e^{2x} \operatorname{cosec} x$, hence

$$e^{2x} \operatorname{cosec} x \cdot p = \int e^{2x} \operatorname{cosec} x \sin x \, dx = \int e^{2x} dx = \frac{1}{2} e^{2x} + c_1$$

Or $p = \frac{1}{2} \sin x + c_1 e^{-2x} \sin x$. Integrating, we obtain

$$v = \int p \, dx = -\tfrac{1}{2} \cos x + c_1 \frac{e^{-2x}}{5} (-\cos x - 2\sin x) + c_2 .$$

Finally, the general solution of the differential equation is

$y = A e^x + B e^{-x} (2\sin x + \cos x) - \frac{1}{2} e^x \cos x$. □

Case II: Normal form

Consider again Equation (4)

$$v''+\frac{1}{u}(2u'+Pu)v'+\frac{1}{u}(u''+Pu'+Qu)v=\frac{f(x)}{u}$$

If we choose the function u such that $2u'+Pu=0$, then

$$u=e^{-\int\frac{P}{2}dx},\ u'=-\frac{P}{2}e^{-\int\frac{P}{2}dx} \text{ and}$$

$$u''=\left(\frac{P^2}{4}-\frac{P'}{2}\right)e^{-\int\frac{P}{2}dx}$$

and Equation (4) reduces to

$$v''+\left\{Q-\left[\left(\frac{P}{2}\right)^2+\frac{d}{dx}\left(\frac{P}{2}\right)\right]\right\}v=f(x)e^{-\int\frac{P}{2}dx} \tag{12}$$

This equation is called the ***normal form*** or the ***canonical form***. It can be solved under certain conditions, namely:

1. If the quantity $Q-\left[\left(\frac{P}{2}\right)^2+\frac{d}{dx}\left(\frac{P}{2}\right)\right]$ is a constant, then the differential equation for v reduces to one with constant coefficients.
2. If the quantity $Q-\left[\left(\frac{P}{2}\right)^2+\frac{d}{dx}\left(\frac{P}{2}\right)\right]=\frac{\text{constant}}{x^2}$, then the differential equation for v reduces to Euler equation.
3. If the quantity $Q-\left[\left(\frac{P}{2}\right)^2+\frac{d}{dx}\left(\frac{P}{2}\right)\right]=\frac{\text{constant}}{(ax+b)^2}$, then the differential equation for v reduces to a variation of Euler equation known as Legendre linear equation.
4. If none of the above, try another method!

***Example* 1**: Solve the equation $4x^2y''+4xy'+(x^2-1)y=0$.

Solution: Here $P=\frac{1}{x}$ and $Q=\frac{x^2-1}{4x^2}$, then

$$Q-\left[\left(\frac{P}{2}\right)^2+\frac{d}{dx}\left(\frac{P}{2}\right)\right]=\frac{x^2-1}{4x^2}-\frac{1}{4x^2}+\frac{1}{2x^2}=\frac{1}{4}.$$

The equation for v is $v'' + \frac{1}{4}v = 0$.

The solution v is $v = a\cos\frac{x}{2} + b\sin\frac{x}{2}$,

hence, the general solution is given by

$$y = v\, e^{-\int \frac{P}{2}dx} = \frac{1}{\sqrt{x}}\left(a\cos\frac{x}{2} + b\sin\frac{x}{2}\right). \qquad \square$$

***Example* 2**: Put the equation $y'' + (2\sin x)y' + (\sin^2 x + \cos x - \frac{2}{x^2})y = e^{\cos x}$ in normal form, then find the general solution.

Solution: Let $y = uv$ then $y' = uv' + vu'$ and $y'' = uv'' + 2u'v' + vu''$.

Substituting in the differential equation and rearranging, we get

$$v'' + \frac{1}{u}[2u' + (2\sin x)u]v' + \frac{1}{u}[u'' + (2\sin x)u' + (\sin^2 x + \cos x - \frac{2}{x^2})u]v = e^{\cos x}$$

We choose u such that $2u' + (2\sin x)u = 0$,

then $u = e^{-\int \sin x\, dx} = e^{\cos x}$ and

$u' = -e^{\cos x}\sin x$ and $u'' = e^{\cos x}(\sin^2 x - \cos x)$.

Substituting in the equation for v, we obtain $v'' - \frac{2}{x^2}v = 1$.

This is an Euler equation and its solution can be found to be

$v = ax^2 + bx^{-1} + \frac{1}{3}x^2\ln x$.

The reader can easily verify this. Finally, the general solution will be

$$y = e^{\cos x}\left(ax^2 + bx^{-1} + \frac{1}{3}x^2\ln x\right). \qquad \square$$

Exercise 4.1

a. Show that e^{2x} is a solution of the homogeneous equation: $(x-2)y''-(4x-7)y'+(4x-6)y=0$, then find the general solution.

Ans: $y=ae^{2x}(x-2)^2+be^{2x}$

b. Show that e^x is a solution of the equation: $xy''-(x+2)y'+2y=0$, then find the general solution. **Ans:** $y=ae^x+b(x^2+2x+2)$

c. Solve the equation $y''+[1+2x\cot x-2/x^2]y=x\cos x$ if $(\sin x)/x$ is a part of the complementary function.

d. Solve the following differential equations:

1. $(1+x^2)y''-2xy'+2y=2$ **Ans:** $u=x$, $y=ax+b(x^2-1)+x^2$
2. $x^2y''-2x(1+x)y'+2(1+x)y=x^3$ **Ans:** $y=ax\,e^{2x}+bx-x^2/2$
3. $x^2y''-(x^2+2x)y'+(2+x)y=x^3e^x$ **Ans:** $y=ax\,e^x+bx+x(x-1)e^x$
4. $(x-x^2)y''-(1-2x)y'+(1-3x+x^2)y=(1-x)^2$

   Ans: $y=ax^2e^x+be^x-x$

e. Put the following equations in normal form then find all solutions:

1. $y''-\frac{2}{x}y'+(1+\frac{2}{x^2})y=x\,e^x$ **Ans:** $y=ax\cos x+bx\sin x-\frac{1}{2}x\,e^x$
2. $y''-2xy'+(x^2+2)y=e^{\frac{x^2+2x}{2}}$

   Ans: $y=e^{x^2/2}(a\cos\sqrt{3}x+b\sin\sqrt{3}x)+\frac{1}{4}e^{\frac{x^2+2x}{2}}$
3. $y''-\frac{3}{x}y'+\frac{3}{x^2}y=2x-1$ **Ans:** $y=ax+bx^3+x^3\ln x+x^2$
4. $y''-4xy'+4x^2y=x\,e^{x^2}$ **Ans:** $y=e^{x^2}(a\cos\sqrt{2}x+b\sin\sqrt{2}x)+\frac{x\,e^{x^2}}{2}$
5. $y''-(2\tan x)y'-10y=0$ **Ans:** $y=\sec x\,(ae^{3x}+be^{-3x})$
6. $y''+(2\tan x)y'-5y=e^x\sec x$ **Ans:** $y=\sec x\,(ae^{2x}+be^{-2x}-\frac{1}{3}e^x)$
7. $xy''+2y'+xy=\sin 2x$ **Ans:** $y=(a\cos x+b\sin x-\frac{1}{3}\sin 2x)/x$
8. $y''-4xy'+(4x^2-3)y=e^{x^2}$ **Ans:** $y=e^{x^2}(ae^x+be^{-x}-1)$

4.3. Changing the Independent Variable

Consider the linear second order differential equation

$$\frac{d^2y}{dx^2}+P\frac{dy}{dx}+Qy=R$$

where P, Q and R are function in x. Changing the independent variable x to t such that $t=f(x)$, we have

$$\frac{dy}{dx}=\frac{dy}{dt}\cdot\frac{dt}{dx}, \text{ and}$$

$$\frac{d^2y}{dx^2}=\frac{d}{dx}\left(\frac{dy}{dt}\cdot\frac{dt}{dx}\right)=\left(\frac{dy}{dx}\right)^2\cdot\frac{d^2y}{dt^2}+\frac{d^2t}{dx^2}\cdot\frac{dy}{dt}$$

Substituting in the differential equation, we get

$$\left(\frac{dt}{dx}\right)^2\frac{d^2y}{dt^2}+\frac{d^2t}{dx^2}\frac{dy}{dt}+P\frac{dt}{dx}\frac{dy}{dt}+Qy=R$$

Dividing by $\left(\frac{dt}{dx}\right)^2$, we get

$$\frac{d^2y}{dt^2}+\frac{\frac{d^2t}{dx^2}+P\frac{dt}{dx}}{\left(\frac{dt}{dx}\right)^2}\frac{dy}{dt}+\frac{Q}{\left(\frac{dt}{dx}\right)^2}y=\frac{R}{\left(\frac{dt}{dx}\right)^2}$$

or

$$\frac{d^2y}{dt^2}+P_1\frac{dy}{dt}+Q_1y=R_1$$

where $P_1=\frac{\frac{d^2t}{dx^2}+P\frac{dt}{dx}}{\left(\frac{dt}{dx}\right)^2}$, $Q_1=\frac{Q}{\left(\frac{dt}{dx}\right)^2}$ and $R_1=\frac{R}{\left(\frac{dt}{dx}\right)^2}$

We can see that P_1, Q_1 and R_1 are functions in x. They can be converted to functions in t using the transformation $t=f(x)$.

Now, ***if by equating Q_1 to a constant, P_1 also becomes a constant***, we get a linear equation with constant coefficients that can be solved using known techniques. The following examples illustrate the procedure.

***Example* 1**: Solve the equation $y''+\frac{2x}{1+x^2}y'+\frac{4}{(1+x^2)^2}y=0$.

Solution: We have $P=\frac{2x}{1+x^2}$, $Q=\frac{4}{(1+x^2)^2}$ and $R=0$.

Let us choose t such that $\left(\frac{dt}{dx}\right)^2=\frac{4}{(1+x^2)^2}$, then

$\frac{dt}{dx}=\frac{2}{1+x^2}$ and $\frac{d^2t}{dx^2}=-\frac{4}{(1+x^2)^2}$,

and on integrating $t=2\tan^{-1}x$. From this we have

$$P_1=\frac{-\frac{4x}{(1+x^2)^2}+\frac{2x}{1+x^2}\cdot\frac{2}{1+x^2}}{\frac{4}{(1+x^2)^2}}=0,\ Q_1=1 \text{ and } R_1=0.$$

The differential equation reduces to $\frac{d^2y}{dt^2}+y=0$ whose solution is $y=A\cos t+B\sin t$. Then the general solution of the given equation is $y=A\cos(2\tan^{-1}x)+B\sin(2\tan^{-1}x)$. □

***Example* 2:** Solve the equation $xy''-y'+4x^3y=x^5$.

Solution: Here we have $P=-\frac{1}{x}$, $Q=4x^2$ and $R=x^4$.

We choose t such that

$\left(\frac{dt}{dx}\right)^2=4x^2$ or $\frac{dt}{dx}=2x$, $\frac{d^2t}{dx^2}=2$ and $t=x^2$

From this we have

$$P_1=\frac{\frac{d^2t}{dx^2}+P\frac{dt}{dx}}{\left(\frac{dt}{dx}\right)^2}=\frac{2-\frac{1}{x}\cdot 2x}{4x^2}=0,\ Q_1=\frac{4x^2}{4x^2}=1$$

and $R_1=\frac{1}{4}x^2=\frac{1}{4}t$.

The differential equation becomes $\frac{d^2y}{dt^2}+y=\frac{1}{4}t$, whose solution can be found to be

$$y = A\cos t + B\sin t + \frac{1}{4}t, \text{ or } y = A\cos x^2 + B\sin x^2 + \frac{1}{4}x^2. \quad \square$$

Exercise 4.2

Changing the independent variable by using appropriate transformation, solve the following differential equations:

1. $y'' + \cot x \cdot y' + 4\operatorname{cosec}^2 x \cdot y = 0$

   Ans: $y = a\cos\left[2\ln\tan\left(\frac{x}{2}\right)\right] + b\sin\left[2\ln\tan\left(\frac{x}{2}\right)\right]$

2. $y'' + \frac{3}{x}\cdot y' + \frac{1}{x^6}\cdot y = \frac{1}{x^8}$ **Ans:** $y = a\cos\left(\frac{1}{2x^2}\right) + b\sin\left(\frac{1}{2x^2}\right) + \frac{1}{x^2}$

3. $y'' + \tan x \cdot y' - 2\cos^2 x \cdot y = 2\cos^4 x$

   Ans: $y = ae^{\sqrt{2}\sin x} + be^{-\sqrt{2}\sin x} + \sin^2 x$

4. $xy'' - y' - 4x^3 y = 8x^3 \sin x^2$ **Ans:** $y = ae^{x^2} + be^{-x^2} - \sin x^2$

5. $y'' + \frac{2}{x}\cdot y' + \frac{1}{x^4}\cdot y = 0$ **Ans:** $y = a\cos\left(\frac{1}{x}\right) + b\sin\left(\frac{1}{x}\right)$

6. $y'' + \tan x \cdot y' + \cos^2 x \cdot y = 0$ **Ans:** $y = a\cos(\sin x) + b\sin(\sin x)$

7. $y'' - \cot x \cdot y' - \sin^2 x \cdot y = \sin^2 x \cos x$

   Ans: $y = ae^{\cos x} + be^{-\cos x} - \cos x$

8. $y'' + (3\sin x - \cot x)\cdot y' + 2\sin^2 x \cdot y = \sin^2 x \cdot e^{-\cos x}$

   Ans: $y = ae^{\cos x} + be^{2\cos x} + \frac{1}{6}e^{-\cos x}$

9. $xy'' + (4x^2 - 1)y' + 4x^3 y = 2x^3$

   Ans: $y = e^{-x^2}(a\cos x^2 + b\sin x^2) + \frac{1}{2}$

10. $y'' + (\tan x - 3\cos x)y' + 2\cos^2 x \cdot y = \cos^4 x$

    Ans: $y = ae^{\sin x} + be^{-\sin x} - \frac{1}{2}\sin^2 x - \frac{3}{2}\sin x - \frac{5}{4}$

4.4. Exact Equations

Consider the linear second order equation with variable coefficients

$$[a_0(x)D^2 + a_1(x)D + a_2(x)]y = f(x) \tag{13}$$

The equation is called ***exact*** if it can be put in the form

$$D[b_0(x)D + b_1(x)]y = f(x) \tag{14}$$

To find the condition for exactness in term of the coefficients a_0, a_1 *and* a_2 we say, dropping the argument x for simplicity,

$$D[b_0D + b_1]y = [b_0D^2 + (b_0' + b_1)D + b_1']y = f(x) \tag{15}$$

Comparing Equation (15) with Equation (13), we must have

$$a_0 = b_0, \quad a_1 = b_0' + b_1 \quad \text{and} \quad a_2 = b_1' \tag{16}$$

Eliminating b_0 and b_1 from these three equations, we obtain

$$a_0'' - a_1' + a_2 = 0 \tag{17}$$

This is the necessary condition for exactness, it can be shown that the condition is also sufficient. The coefficient functions are then

$$b_0 = a_0 \quad \text{and} \quad b_1 = a_1 - a_0'$$

and Equation (14) becomes

$$D[a_0D + (a_1 - a_0')]y = f(x).$$

Integrating this last equation, we get

$$[a_0D + (a_1 - a_0')]y = \int f(x)dx + c_1,$$

which is a first order differential equation and is solved by usual means.

We have seen that, if the differential equation is exact, we reduce its order by one by performing one integration step. The idea for exactness can be extended the higher order linear differential equations. For example, for a third order equation, the condition for exactness can be found to be

$$a_0''' - a_1'' + a_2' - a_3 = 0$$

and for an nth order equation,

$$a_0^{(n)} - a_1^{(n-1)} + \cdots + (-1)^n a_n = 0, \quad n \geq 2$$

The following examples demonstrate the method to solve exact second order differential equations.

***Example* 1**: Solve the equation $(1-x^2)y''-3xy'-y=1$.

Solution: The variable coefficients are $a_0=1-x^2$, $a_1=-3x$ and $a_2=-1$.

Then $a_0''-a_1'+a_2=-2+3-1=0$.

Hence the equation is exact. It can be put in the form

$D[(1-x^2)D-x]y=1$.

Integrating, we get $[(1-x^2)D-x]y=x+c_1$

or $\dfrac{dy}{dx}-\dfrac{x}{1-x^2}y=\dfrac{x+c_1}{1-x^2}$.

This is a linear first order equation, and the integrating factor is

$\mu=e^{-\int\frac{x\,dx}{1-x^2}}=\sqrt{1-x^2}$ and $\sqrt{1-x^2}\,y=\int\sqrt{1-x^2}\cdot\dfrac{x+c_1}{1-x^2}dx$.

The general solution becomes $y=c_1\dfrac{\sin^{-1}x}{\sqrt{1-x^2}}+c_2\dfrac{1}{\sqrt{1-x^2}}-1$. □

***Example* 2**: Solve the equation $xy''+(x+2)y'+y=0$.

Solution: Here $a_0=x$, $a_1=x+2$ and $a_2=1$.

Then $a_0''-a_1'+a_2=0-1+1=0$.

Hence the equation is exact. It can be put in the form

$D[xD-(x+1)]y=0$, then $[xD-(x+1)]y=c_1$

or $\dfrac{dy}{dx}+\dfrac{x+1}{x}y=\dfrac{c_1}{x}$.

This is a linear first order equation, and its integrating factor is

$\mu=e^{-\int\left(\frac{1}{x}+1\right)dx}=x\,e^x$

and the general solution is $y=c_1x^{-1}+c_2x^{-1}e^{-x}$. □

Note: Suppose that Equation (13) is not exact. Is there a way to make it exact? Let us see. If the coefficient functions a_0, a_1 and a_2 are algebraic functions and not transcendental, then, in such cases, we can multiply the whole equation by x^m, an integrating factor, and choose m so that the resultant equation becomes exact, easy!. Consider the following example.

***Example* 3**: Solve the differential equation $\sqrt{x}\, y'' + 2xy' + 3y = x$.

Solution: Here we have $a_0 = \sqrt{x}$, $a_1 = 2x$ and $a_2 = 3$.

Hence $a_0'' - a_1' + a_2 \neq 0$, and the equation is not exact. Multiplying the equation by the integrating factor x^m , we get

$$x^{m+1/2} y'' + 2x^{m+1} y' + 3x^m y = x^{m+1} .$$

Now, $a_0 = x^{m+1/2}$, $a_1 = 2x^{m+1}$ and $a_2 = x^m$. Applying the condition for exactness, we must have

$$(m + 1/2)(m - 1/2)x^{m-3/2} - 2(m - 1/2)x^m \equiv 0 .$$

It is clear that the value $m = 1/2$ will make the equation exact. The taking this value the equation becomes

$$xy'' + 2x\sqrt{x}\, y' + 3\sqrt{x}\, y = x\sqrt{x}$$

Here, $a_0 = x$, $a_1 = 2x\sqrt{x}$ and $a_2 = 3\sqrt{x}$, and the equation reduces to

$$D[a_0 D + (a_1 - a_0')]y = D[xD + (2x\sqrt{x} - 1)]y = x\sqrt{x}$$

Integrating once, we obtain

$$[xD + (2x\sqrt{x} - 1)]y = \tfrac{2}{5}x^2\sqrt{x} + c_1$$

or $$\frac{dy}{dx} + \frac{2x\sqrt{x} - 1}{x} y = \tfrac{2}{5}x\sqrt{x} + \frac{c_1}{x} .$$

This is a linear first order equation whose solution is found to be

$$\frac{y}{x} e^{4x\sqrt{x}/3} = \tfrac{1}{5} e^{4x\sqrt{x}/3} + c_1 \int \frac{e^{4x\sqrt{x}/3}}{x^2} dx + c_2 .$$ □

Exercise 4.3

a. Show that the following differential equations are exact and find all solutions:

1. $xy''+(3x+1)y'+3y=0$
2. $2x^2y''+5xy'+y=0$
3. $4xy''+x^2y'+2xy=0$
4. $2x^2y''+8xy'+4y=0$
5. $x\,e^{2x}\,y''+(2+4x)e^{2x}\,y'+4(1+x)e^{2x}\,y=0$
6. $xy''+(1-x)y'-y=e^x$
7. $x^2y''+3xy'+y=\dfrac{1}{(1-x)^2}$

b. Find an integrating factor of the form x^m to make the following equations exact and find all solutions:

1. $x^5y''+3x^3y'+(3-6x)x^2y=x^4+2x-5$

$$\textbf{Ans: } \frac{y}{x^3}e^{-3/x}=\int\left(\frac{1}{3}+\frac{2}{x^3}\ln x+\frac{5}{x^4}+\frac{c_1}{x^3}\right)\cdot\frac{1}{x^3}e^{-3/x}\,dx+c_2$$

2. $2x^2(x+1)y''+x(7x+3)y'-3y=x^2$

$$\textbf{Ans: } 5(x+1)=\tfrac{5}{7}x^2+c_1x-c_2x^{-3/2}$$

3. $x^4y''+x^2(x-1)y'+xy=x^3-4$

$$\textbf{Ans: } y\,e^{1/x}=\int\left(1-\frac{c_1}{x}\right)e^{1/x}\,dx-2e^{1/x}\left(\frac{1}{x}-1\right)+c_2$$

4.5. Operator Factorization

In the linear second order differential equation

$$[a_0(x)D^2 + a_1(x)D + a_2(x)]y = f(x) \tag{18}$$

we can sometimes factor the second order operator $[a_0D^2 + a_1D + a_2]$ into the product of two first order operators, i.e., (we have dropped the argument x for simplicity),

$$a_0D^2 + a_1D + a_2 = (b_0D + b_1)(b_2D + b_3)$$

where b_0, b_1, b_2 and b_3 can be functions in x. If this is the case, then

$$(b_0D + b_1)(b_2D + b_3)y = f(x)$$

If we let
$$(b_2D + b_3)y = w \tag{19}$$

then
$$(b_0D + b_1)w = f(x) \tag{20}$$

Clearly Equations (19) and (20) are two first order differential equations the first in y and the second in w. Solving (20) for w, then (19) for y, we obtain the general solution of the given differential equation.

Note: The factors are not commutative since they involve functions of x directly, so correct order should be strictly maintained.

***Example* 1**: Solve the differential equation $[xD^2 + (2+x)D + 1]y = x$.

Solution: By operator factorization, we obtain $(D+1)(xD+1)y = x$.

Let $(xD+1)y = w$ then $(D+1)w = x$.

The solution w is $w = c_1e^{-x} + x - 1$, and the equation for y becomes $\dfrac{dy}{dx} + \dfrac{1}{x}y = c_1\dfrac{e^x}{x} + 1 - \dfrac{1}{x}$. This is a first order linear differential equation. The integrating factor μ is $\mu = e^{\int \frac{1}{x}dx} = x$.

Hence $xy = \displaystyle\int (c_1e^x + x - 1)dx = c_1e^x + \frac{x^2}{2} - x + c_2$

Then the general solution is $y = c_1x^{-1}e^x + c_2x^{-1} + \frac{1}{2}x - 1$. □

***Example* 2**: Using operator factorization, solve the differential equation:

$$[(x+2)D^2 - (2x+5)D + 2]y = (x+1)e^x .$$

Solution: The operator on the left hand side can be factored as follows:

$$(x+2)D^2-(2x+5)D+2=(x+2)D^2-2(x+2)D-(D-2)$$

$$=(x+2)D(D-2)-(D-2)=[(x+2)D-1](D-2).$$

Then the differential equation is written as

$[(x+2)D-1](D-2)y=(x+1)e^x$.

Letting $(D-2)y=w$, then $[(x+2)D-1]w=(x+1)e^x$.

This is a linear equation that can be written in the usual form

$\frac{dw}{dx}-\frac{1}{x+2}w=\frac{x+1}{x+2}e^x$, whose integrating factor is

$\mu=e^{-\int\frac{dx}{x+2}}=\frac{1}{x+2}$, and its solution is found to be

$w=c_1(x+2)+e^x$.

Now, we have $(D-2)y=c_1(x+2)+e^x$.

Again, this is a linear equation in *y*: $\frac{dy}{dx}-2y=c_1(x+2)+e^x$,

whose solution is $y=c_2e^{2x}-\frac{1}{4}c_1(2x+5)-e^x$. □

Exercise 4.4

Solve the following differential equation by operator factorization:

1. $[(x+1)D^2+(x-1)D-2]y=0$ **Ans**: $y=a(x^2+1)+be^{-x}$

2. $[(x+3)D^2-(2x+7)D+2]y=(x+3)^3e^x$

   Ans: $y=a(2x+7)+be^{2x}-xe^x-4e^x$

3. $[(x+1)D^2-(3x+4)D+3]y=(3x+2)^2e^{3x}$

   Ans: $y=a(3x+4)+be^{3x}+xe^{3x}$

4. $[(x+2)D^2-(2x+5)D+2]y=(x+1)e^x$

   Ans: $y=ae^{2x}+b(2x+5)-e^x$

5. $[xD^2+(x-2)D-2]y=x^3$

   Ans: $y=x^3+(a-3)x^2+(6-2a)x+2(a-3)+be^{-x}$

4.6. The Method of Variation of Parameters

In the previous chapter, we gave full details on the method of variation of parameters. We mentioned that the method is applicable to linear differential equations with constant coefficients as well as variable coefficients. The assumption that the two linearly independent solutions of the related homogeneous equation be known is in force. For the sake of clarity, we restate the steps for the solution.

Step 1: Solve for the complementary function $y_{CF} = a_1 y_1 + a_2 y_2$.

Step 2: Compute the Wronskian $W = y_1 y_2' - y_2 y_1'$.

Step 3: Assume a particular solution of the form $y_{PI} = v_1 y_1 + v_2 y_2$.

Step 4: Compute $v_1 = -\int \frac{y_2 f(x)}{W} dx + a_1$ and $v_2 = \int \frac{y_1 f(x)}{W} dx + a_2$.

Step 5: The general solution is $y = y_{CF} + y_{PI}$.

We give here two examples for linear differential equations with variable coefficients.

***Example* 1**: Using the method of variation of parameters find a solution for

$$x^2 y'' - 4y' + 4y = x^2(x^2+1).$$

Solution: We first find the solution of the related homogeneous equation:

$$y'' - \frac{4}{x} y' + \frac{4}{x^2} y = 0.$$

This is clearly Euler's homogeneous equation. The solution can be found to be $y_{CF} = ax + bx^4$.

Then, the two linearly independent solutions of the related homogeneous equation are $y_1 = x$ and $y_2 = x^4$.

The Wronskian is $W = \begin{vmatrix} x & x^4 \\ 1 & 4x^3 \end{vmatrix} = 4x^4 - x^4 = 3x^4$.

If we assume that the particular solution is of the form

$y = xv_1 + x^4 v_2$, then,

$$v_1 = -\int \frac{y_2 f(x) dx}{W} = -\int \frac{x^4(x^2+1)dx}{3x^4} = -\frac{1}{9}x^3 - \frac{1}{3}x$$

and $v_2 = \int \frac{y_1 f(x) dx}{W} = \int \frac{x(x^2+1)dx}{3x^4} = \frac{1}{3}\ln x - \frac{1}{6}x^{-2}$

We notice that $f(x)$ is taken on the assumption that the coefficient of y'' in the original differential equation is unity, i.e., we have divided by x^2 first.

The particular solution is $y_{PI} = -\frac{1}{9}x^4 - \frac{1}{2}x^2 + \frac{1}{3}x^4 \ln x$.

And the general solution will be

$y = ax + bx^4 - \frac{1}{9}x^4 - \frac{1}{2}x^2 + \frac{1}{3}x^4 \ln x$. □

***Example* 2**: Verify that e^x and $1/x$ are solutions of the related homogeneous equation of $x(x+1)y'' + (2-x^2)y' - (2+x)y = (x+1)^2$; show that they are linearlyindependent, then use the method of variation of parameters to find the general solution.

Solution: The related homogeneous equation is

$x(x+1)y'' + (2-x^2)y' - (2+x)y = 0$

For $y_1 = e^x$, we have $y_1' = e^x$ and $y_1'' = e^x$; substituting we get $e^x[x(x+1) + (2-x^2) - (2+x)] = 0$; hence, $y_1 = e^x$ is a solution.

For $y_2 = x^{-1}$, we have $y_2' = -x^{-2}$ and $y_2'' = 2x^{-3}$, then substituting we get $x(x+1)\frac{2}{x^3} + (2-x^2)(-\frac{1}{x^2}) - (2+x)\frac{1}{x} = 0$;

hence, $y_2 = x^{-1}$ is a solution. The Wronskian is given by

$$W = \begin{vmatrix} e^x & 1/x \\ e^x & -1/x^2 \end{vmatrix} = -e^x \frac{1+x}{x^2} \neq 0, x \neq 0$$

Then the two solutions are linearly independent, and the complementary function is $y_{CF} = ae^x + bx^{-1}$.

For the particular solution, if we assume that

$y = e^x v_1 + x^{-1} v_2$, then,

$$v_1 = -\int \frac{y_2 f(x)dx}{W} = \int \frac{1}{x} \cdot \frac{x+1}{x} \cdot \frac{x^2}{1+x} e^{-x} dx = -e^{-x}$$

and $v_2 = \int \frac{y_1 f(x)dx}{W} = -\int e^x \cdot \frac{x+1}{x} \cdot \frac{x^2}{x+1} \cdot e^{-x} dx = -\frac{x^2}{2}$

and the particular solution becomes $y_{PI} = -\frac{1}{2}(x+2)$.

The general solution will be $y = ae^x + bx^{-1} - \frac{1}{2}(x+2)$. □

Exercise 4.5

a. Verify that the functions between brackets are two linearly independent solutions of the related homogeneous equation of the following equations, then use the method of variation of parameters to find the general solution:

1. $x^2y''-2xy'+2y=x\ln x\ ,\ (x, x^2)$

Ans: $y=ax+bx^2-\frac{1}{2}x(\ln x)^2-x(\ln x+1)$

2. $xy''-(2x^2+1)y'=x^5e^{x^2}\ ,\ (1, e^{x^2})$

Ans: $y=a+be^{x^2}+\frac{1}{4}(1-x^2+\frac{1}{2}x^4)e^{x^2}$

3. $(x^2-2x)y''+2(1-x)y'+2y=6(x^2-2x)^2\ ,\ (x^2, x-1)$

Ans: $y=ax^2+b(x-1)+x^4-4x^3$

4. $x(x+1)y''+(2-x^2)y'-(2+x)y=(x+1)^2\ \ (e^x, x^{-1})$

Ans: $y=ae^x+bx^{-1}-\frac{1}{2}(x+2)$

5. $(1-x)y''+xy'-y=2(x-1)^2e^{-x}\ \ (e^x, x)$

Ans: $y=ax+be^x+e^{-x}(1/2-x)$

6. $(x-1)y''-xy'+y=(x-1)^2\ \ (e^x, x)$

Ans: $y=ae^x+bx-(x^2+x+1)$

b. Solve the equation $x^2y''+xy'-y=x^2e^x$.

Ans: $y=ax+bx^{-1}+e^x(1-x^{-1})$

c. Solve the first order linear differential equation $y'+P(x)y=Q(x)$ by first solving the related homogeneous equation, and then obtaining a particular solution by the method of variation of parameters.

d. Consider the second order differential equation with constant coefficients $y''+by'+cy=f(x)$. If the roots λ_1 and λ_2 of the auxiliary equation are distinct, apply the method of variation of parameters to show that the particular solution is

$$y_{PI}=\frac{e^{\lambda_1 x}}{\lambda_1-\lambda_2}\int e^{-\lambda_1 x}f(x)dx+\frac{e^{\lambda_2 x}}{\lambda_2-\lambda_1}\int e^{-\lambda_2 x}f(x)dx$$

Chapter Five

Series Solutions of Differential Equations

$$y = \sum_{k=0}^{\infty} c_k (x - x_0)^{k+\alpha}, \quad c_0 \neq 0$$

Chapter 5.

Series Solutions of Differential Equations

5.1. Introduction

It may happen that a differential equation cannot be solved by any of the preceding classes of methods designed to find the solutions for some particular form of equations. In such situations we must find other means to express the solutions. These methods are related to power series. There are several ways in which series are used in differential equations. We consider here methods designed to obtain these power series solutions. We restrict ourselves to second order differential equations although the methods presented apply to higher order equations.

To get the idea, consider the differential equation

$$y'' - 5y' + 6y = 0$$

the solution of this equation can be found by usual means as

$$y = ae^{2x} + be^{3x}$$

If we expand the functions e^{2x} and e^{3x} in power series about $x = 0$, then the solution of the differential equation can be written in power series form as

$$y = a\left(1 + 2x + \frac{(2x)^2}{2!} + \frac{(2x)^3}{3!} + \cdots\right) + b\left(1 + 3x + \frac{(3x)^2}{2!} + \frac{(3x)^3}{3!} + \cdots\right)$$

From this, we see that the general solution of a second order differential equation may be expressed as a linear combination of two infinite series. This suggests that we try solutions for differential equations in terms of infinite power series.

5.2. Taylor's Series Method

If $f(x)$ is defined in some interval $a < x < b$ and if x_0 is a point in this interval and if all derivatives of $f(x)$ exist at x_0, then Taylor's series of $f(x)$ is

$$f(x) = f(x_0) + (x - x_0)f'(x_0) + \frac{(x - x_0)^2}{2!}f''(x_0) + \frac{(x - x_0)^3}{3!}f'''(x_0) + \cdots + \frac{(x - x_0)^n}{n!}f^{(n)}(x_0) + \cdots$$

If $x_0 = 0$, we obtain MacLaurin's (***Colin Maclaurin*** (1698-1746)) series

$$f(x)=f(0)+xf'(0)+\frac{x^2}{2!}f''(0)+\frac{x^3}{3!}f'''(0)+\cdots+\frac{x^n}{n!}f^{(n)}(0)+\cdots.$$

The series will converge to $f(x)$ for all x in an interval with x_0 as midpoint under appropriate hypotheses.

Examples of convergent series are

$$e^x=1+x+\frac{x^2}{2!}+\cdots+\frac{x^n}{n!}+\cdots \text{ (convergent } \forall x)$$

$$\sin x=x-\frac{x^3}{3!}+\frac{x^5}{5!}-\frac{x^7}{7!}+\cdots+\frac{(-1)^n x^{2n+1}}{(2n+1)!}+\cdots \text{ (convergent } \forall x)$$

$$\frac{1}{1-x}=1+x+x^2+x^3+\cdots+x^n+\cdots \; (\textit{convergent for } |x|<1)$$

If a series converges for all x, it is convenient to say that ***the radius of convergence is infinity***. If the series diverges for all $x\neq 0$, we say that the radius of convergence is zero. Each power series has a radius of convergence R and the series is convergent when $|x|<R$.

Consider the second order differential equation

$$y''=F(x,y,y') \tag{1}$$

It is reasonable to expect a solution of (1), that is a power series in x, to contain two arbitrary constants. We can assign the values of y and y' at $x=0$ to be these two arbitrary constants, i.e., at $x=0$, $y(0)=A$ and $y'(0)=B$.

So, $y''(0)$ can be computed directly from Equation (1) in terms of A and B. Differentiating Equation (1) with respect to x, we get

$$y'''=\frac{d}{dx}F(x,y,y') \tag{2}$$

Again $y'''(0)$ can be computed from Equation (2) in terms of A and B. This process can be continued to compute higher derivatives of y at $x=0$. Then using MacLaurin's expansion, the solution y is

$$y(x)=y(0)+xy'(0)+\frac{x^2}{2!}y''(0)+\frac{x^3}{3!}y'''(0)+\cdots+\frac{x^n}{n!}y^{(n)}(0)+\cdots \tag{3}$$

The series in Equation (3) will converge to the value of $y(x)$ is some interval about $x=0$, if $y(x)$ is well behaved at and near $x=0$.

Note: If the initial conditions are not at $x = 0$ but at $x = a$, then we use Taylor's expansion instead.

We give some examples for the application of Taylor's or MacLaurin's expansions to find series solutions for differential equations.

***Example* 1**: Find the first four non-zero terms in a power series solution for the differential equation: $y' = x^2 - y^2$, when $x = 1, y = 1$.

Solution: We have

$$y(1) = 1$$

$$y' = x^2 - y^2 \qquad y'(1) = 0$$

$$y'' = 2x - 2yy' \qquad y''(1) = 2$$

$$y''' = 2 - 2y'^2 - 2yy'' \qquad y'''(1) = -2$$

$$y^{\text{iv}} = -6y'y'' - 2yy''' \qquad y^{\text{iv}}(1) = 4$$

Then the power series solution is

$$y = 1 + \frac{2(x-1)^2}{2!} - \frac{2(x-1)^3}{3!} + \frac{4(x-1)^4}{4!} + \cdots. \qquad \square$$

This example reveals the following points:

1. Can we write the general term for the series? sometimes tricky! and most of the time impossible.
2. Does the series converge to $y(x)$? If we can write the general term, we can use the ratio test to obtain the values of x for which the series is convergent. Hopefully, $x = 0$ is inside the interval of convergence.
3. If the series is convergent, how rapidly is this convergence?
4. If we take only the first k terms of the series as an approximation to $y(x)$, how big can the error in approximation be?

These are some questions to be asked, but in the current context, we will not be able to answer some of them.

***Example* 2**: Find the power series solution for the initial value problem:

$$y'' - e^{-x}y' - e^{-x}y^2 + 1 = 0, \text{ when } x = 0,\ y = 1 \text{ and } y' = 1.$$

Solution: Here $y(0) = 1$ and $y'(0) = 1$, and

$$y'' = e^{-x} y' + e^{-x} y^2 - 1 \qquad y''(0) = 1$$

$$y''' = e^{-x}(y'' + 2yy' - y' - y^2) \qquad y'''(0) = 1$$

$$y^{\text{iv}} = e^{-x}(y''' + 2y'^2 + 2yy'' - 2y'' - 4yy' + y^2 + y')$$

$$y^{\text{iv}}(0) = 1$$

Substituting these values in MacLaurin's expansion, we obtain

$$y = 1 + x + \frac{x^2}{2!} + \frac{x^3}{3!} + \cdots + \frac{x^n}{n!} + \cdots$$

The form of this series suggests that $y = e^x$ is the solution of the differential equation, indeed it is. The series in this particular example converges for all finite values of x. □

***Example* 3**: Find the power series solution for the initial value problem:

$$(x-1)y''' + y'' + (x-1)y' + y = 0, \text{ when } x = 0,\ y = y'' = 0 \text{ and } y' = 1.$$

Solution: Here $y(0) = 0$, $y'(0) = 1$ and $y''(0) = 0$.

Solving for y''' then differentiating, we get

$$y''' = -(x-1)^{-1} y'' - y' - (x-1)^{-1} y,$$

$$y^{iv} = -(x-1)^{-1} y''' + [(x-1)^{-2} - 1]y'' - (x-1)^{-1} y' + (x-1)^{-2} y$$

$$y^{v} = -(x-1)^{-1} y^{iv} + [2(x-1)^{-2} - 1]y''' - [2(x-1)^{-3} + (x-1)^{-1}]y'' + 2(x-1)^{-2} y' - 2(x-1)^{-3} y$$

At $x = 0$, we have

$$y'''(0) = -1, \quad y^{iv}(0) = 0 \text{ and } y^{v}(0) = 1.$$

Substituting these values in MacLaurin's expansion, we obtain

$$y = x - \frac{1}{3!}x^3 + \frac{1}{5!}x^5 + \cdots$$

The form of this series shows that $y = \sin x$ is the solution of the differential equation. This series converges for all finite values of x. □

Exercise 5.1

Obtain the first four non-zero terms of the power solution for each of the following initial value problems:

1. $y' = x^2y^2 + 1$, when $x = 1, y = 1$.

   Ans: $y = 1 + 2(x-1) + 3(x-1)^2 + \frac{19}{3}(x-1)^3 + \cdots$

2. $y' = \sin(xy) + x^2$, when $x = 0, y = 3$.

   Ans: $y = 3 + \frac{3}{2}x^2 + \frac{1}{3}x^3 - \frac{3}{4}x^4 + \cdots$

3. $y'' = x^2 - y^2$, when $x = 0, y = 1$ and $y' = 0$.

   Ans: $y = 1 - \frac{1}{2}x^2 + \frac{1}{6}x^4 - \frac{7}{360}x^6 + \cdots$

4. $y''' = xy + yy'$, when $x = 0, y = 0, y' = 1$ and $y'' = 2$.

   Ans: $y = x + x^2 + \frac{1}{24}x^4 + \frac{1}{15}x^5 + \cdots$

5. $y' = x + e^y$, when $x = 0, y = a$.

   Ans: $y = a + e^a x + (1 + e^{2a})\frac{x^2}{2} + (2e^{3a} + e^a)\frac{x^3}{6} + \cdots$

6. $y'' + y = 0$, when $x = 0, y = A$ and $y' = B$.

   Ans: $y = A[1 - \frac{x^2}{2!} + \cdots] + B[x - \frac{x^3}{3!} + \cdots]$

7. $y'^3 + 3xy'^2 + x - y = 0$, when $x = 0$, $y = 1$.

   Ans: $y = 1 + x - \frac{1}{2}x^2 + \frac{5}{18}x^3 + \cdots$

8. $3y^2y' = y^3 - x$, when $x = 0$, $y = 1$.

   Ans: $y = \frac{1}{81}(81 + 27x - 9x^2 + 5x^3 + \cdots)$

5.3. The Method of Undetermined Coefficients

We assume that a series solution of a differential equation takes the form

$$y = c_0 + c_1(x - x_0) + c_2(x - x_0)^2 + \cdots \tag{4}$$

The coefficients c_k, $k = 0, 1, 2, \cdots$ are to be determined so that y satisfies the differential equation. Moreover, the series must be convergent at and near $x = x_0$. In more compact notation the series in Equation (4) can be written as

$$y = \sum_{k=0}^{\infty} c_k (x - x_0)^k \tag{5}$$

The method of undetermined coefficients is based on two facts:

1. If the power series in (5) converges for $|x - x_0| < R$, then the series obtained by differentiating term by term also converges for $|x - x_0| < R$, and represents y'.
2. If the power series in (5) has a sum of zero for $|x - x_0| < R$, then each coefficients c_k, $k = 0, 1, 2, \cdots$ must be zero.

The method of undetermined coefficients is most effective for linear differential equations. We give here three illustrative examples.

***Example* 1**: Find a power series solution for the equation $y'' + xy' + y = 0$ about $x = 0$.

Solution: We start by assuming a solution of the form $y = \sum_{k=0}^{\infty} c_k x^k$.

Then differentiating with respect to *x* twice, we obtain

$$y' = \sum_{k=1}^{\infty} k c_k x^{k-1} = \sum_{k=0}^{\infty} (k+1) c_{k+1} x^k$$

$$y'' = \sum_{k=2}^{\infty} k(k-1) c_k x^{k-2} = \sum_{k=0}^{\infty} (k+1)(k+2) c_{k+2} x^k$$

Substituting in the differential equation, we obtain

$$\sum_{k=0}^{\infty} (k+1)(k+2) c_{k+2} x^k + x \sum_{k=0}^{\infty} (k+1) c_{k+1} x^k + \sum_{k=0}^{\infty} c_k x^k$$

Now, since the sum in the last equation must be identically zero then each coefficient of various powers of *x* must in turn be zero. In particular, equating the coefficients of x^k to zero, we obtain

$$(k+1)(k+2)c_{k+2}+(k+1)c_k=0$$

or $c_{k+2}=-\dfrac{c_k}{k+2},\quad k\geq 0.$

This is called the ***recursion formula*** or the ***recurrence relation*** for the coefficients c_k's. If c_0 and c_1 are kept arbitrary, then all the other c_k's will be in terms of these two arbitrary constants; then substituting for $k=0,\ 1,\ 2,\ \cdots$ in the recursion formula, we obtain

$$c_2=-\frac{c_0}{2}\qquad\qquad c_3=-\frac{c_1}{3}$$

$$c_4=\frac{c_0}{2\cdot 4}\qquad\qquad c_5=\frac{c_1}{3\cdot 5}$$

$$c_6=-\frac{c_0}{2\cdot 4\cdot 6}\qquad\qquad c_7=-\frac{c_1}{3\cdot 5\cdot 7}$$

. . .

$$c_{2n}=\frac{(-1)^n c_0}{2\cdot 4\cdot 6\cdots(2n)}\qquad\qquad c_{2n+1}=\frac{(-1)^n c_1}{3\cdot 5\cdot 7\cdots(2n+1)}$$

And $n=0,\ 1,\ 2,\ \cdots$. The constants c_0 and c_1 are in fact the initial values of y and y' at $x=0$. The series solution can be written as

$$y=c_0\left[1+\sum_{n=1}^{\infty}\frac{(-1)^n x^{2x}}{2\cdot 4\cdot 6\cdots(2n)}\right]+c_1\left[\sum_{n=1}^{\infty}\frac{(-1)^{n+1}x^{2n-1}}{3\cdot 5\cdot 7\cdots(2n-1)}\right]$$

Using the ratio test we can verify that the series converge for all finite x. □

Example 2: Find a power series solution for the differential equation:

$$(x^2-1)y''+x\,y'-y=0 \text{ near } x=0.$$

Solution: Assume a solution of the form $y=\sum_{k=0}^{\infty}c_k x^k$.

Then differentiating with respect to x twice, we obtain

$$y' = \sum_{k=1}^{\infty} k c_k x^{k-1} = \sum_{k=0}^{\infty} (k+1) c_{k+1} x^k$$

$$y'' = \sum_{k=2}^{\infty} k(k-1) c_k x^{k-2} = \sum_{k=0}^{\infty} (k+1)(k+2) c_{k+2} x^k$$

Substituting in the differential equation, we obtain

$$(x^2-1) \sum_{k=0}^{\infty} (k+1)(k+2) c_{k+2} x^k + x \sum_{k=0}^{\infty} (k+1) c_{k+1} x^k - \sum_{k=0}^{\infty} c_k x^k = 0$$

Equating the coefficients of the general power of x (x^k) to zero, we obtain

$$k(k-1)c_k - (k+1)(k+2)c_{k+2} + k c_k - c_k = 0 .$$

Or $c_{k+2} = \frac{k-1}{k+2} c_k$.

This is the ***recurrence relation*** for the coefficients c_k's. If c_0 and c_1 are kept arbitrary, then all the other c_k's will be in terms of these two arbitrary constants; then substituting for $k = 0, 1, 2, \cdots$ in the recurrence relation, we obtain

$$k=0: \quad c_2 = -\frac{1}{2} c_0 \qquad\qquad k=1: \quad c_3 = 0$$

$$k=2: \quad c_4 = \frac{1}{4} c_2 = -\frac{1}{2\cdot 4} c_0 \qquad k=3: \quad c_5 = 0$$

$$k=4: \quad c_6 = \frac{3}{6} c_4 = -\frac{3\cdot 1}{2\cdot 4\cdot 6} c_0 \qquad c_{2n+1} = 0, \quad n \geq 1$$

$$c_{2n} = -\frac{(2n-3)(2n-1)\cdots 3\cdot 1}{2\cdot 4\cdot 6\cdots\cdots 2n} c_0, \quad n \geq 2$$

The general solution is

$$y = A\left\{1 + \frac{1}{2}x^2 + \frac{1}{2\cdot 4}x^4 + \frac{3\cdot 1}{2\cdot 4\cdot 6}x^6 + \cdots\right\} + Bx \text{ , or}$$

$$y = A\left\{1 + \frac{1}{2}x^2 + \sum_{n=2}^{\infty} \frac{1\cdot 3\cdot 5\cdots\cdots(2n-3)}{2\cdot 4\cdot 6\cdots\cdots 2n} x^{2n}\right\} + Bx . \qquad \square$$

***Example* 3**: Find a power series solution for the differential equation:

$$(x^2-1)y''+4x\,y'+2y=0 \text{ near } x=0.$$

Solution: Assume a solution of the form $y=\sum_{k=0}^{\infty} c_k x^k$.

Then differentiating with respect to x twice, we obtain

$$y'=\sum_{k=1}^{\infty} k\,c_k\,x^{k-1}=\sum_{k=0}^{\infty}(k+1)c_{k+1}x^k$$

$$y''=\sum_{k=2}^{\infty} k(k-1)c_k\,x^{k-2}=\sum_{k=0}^{\infty}(k+1)(k+2)c_{k+2}x^k$$

Substituting in the differential equation, we obtain

$$(x^2-1)\sum_{k=0}^{\infty}(k+1)(k+2)c_{k+2}x^k+4x\sum_{k=0}^{\infty}(k+1)c_{k+1}x^k+2\sum_{k=0}^{\infty}c_k x^k=0$$

Equating the coefficients of the general power of x (x^k) to zero, we obtain

$$k(k-1)c_k-(k+1)(k+2)c_{k+2}+4k\,c_k+2c_k=0.$$

Or $c_{k+2}=c_k$.

This is the ***recurrence relation*** for the coefficients c_k's. If c_0 and c_1 are kept arbitrary, then all the other c_k's will be in terms of these two arbitrary constants; then substituting for $k=0, 1, 2, \cdots$ in the recurrence relation, we obtain

$k=0$: $c_2=c_0$ $\qquad$ $k=1$: $c_3=c_1$

$k=2$: $c_4=c_2=c_0$ $\qquad$ $k=3$: $c_5=c_3=c_1$

$c_{2n}=c_0,\ n\ge 1$ $\qquad$ $c_{2n+1}=c_1,\ n\ge 1$

The general solution is

$$y=A\left\{1+x^2+x^4+x^6+\cdots+x^{2n}+\cdots\right\}$$
$$+B\left\{x+x^3+x^5+x^7+\cdots+x^{2n+1}+\cdots\right\},$$

or

$$y=A\left\{\sum_{n=0}^{\infty}x^{2n}\right\}+B\,x\left\{\sum_{n=0}^{\infty}x^{2n}\right\}=\frac{A+Bx}{1-x^2}.$$ □

Note: In the previous examples, the power series solution has been obtained rather in a straightforward manner. This is because the differential equation is "well behaved". We now give some understanding on the behavior of the differential equation.

Definition: A function $g(x)$ is said to be <u>analytic</u> at x_0 if $g(x)$ has a Taylor series expansion which converges to $g(x)$ in some interval about x_0.

For example, $\ln(1+x)$ is analytic at $x=0$. In fact, it has a Taylor series expansion about $x=0$ given by

$$\ln(1+x)=x-\frac{x^2}{2}+\frac{x^3}{3}-\frac{x^4}{4}+\cdots$$

and the interval of convergence is $|x|<1$.

Theorem: If $P(x)$, $Q(x)$ and $f(x)$ are analytic at x_0, then every solution of the linear second order differential equation

$$y''+P(x)y'+Q(x)y=f(x)$$

is also analytic at x_0, and hence is represented by a power series of the form $\sum_{k=0}^{\infty} c_k (x-x_0)^k$. The interval of convergence of the power series solution is at least as large as the smallest of the interval of convergence of $P(x)$, $Q(x)$ and $f(x)$ about x_0.

Looking back at the previous examples, we can see that the coefficients of y and its first and second derivatives are all analytic for all finite values of x. Then, we would expect that the two power series in the solution of the equation will converge for all finite values of x. Moreover, the two power series solutions are linearly independent.

***Example* 4**: Find a power series solution for the following differential equation about $x=0$: $y''+xy'-y=e^{2x}$.

Solution: The coefficient functions x and -1 and $f(x)=e^{2x}$ are all analytic everywhere. Then, we seek a solution of the form $y=\sum_{k=0}^{\infty} c_k x^k$.

Differentiating with respect to x twice, we obtain

$$y'=\sum_{k=1}^{\infty} k c_k x^{k-1}=\sum_{k=0}^{\infty}(k+1)c_{k+1}x^k$$

$$y''=\sum_{k=2}^{\infty} k(k-1)c_k x^{k-2}=\sum_{k=0}^{\infty}(k+1)(k+2)c_{k+2}x^k$$

The function e^{2x} has a power series expansion about $x = 0$ given by

$$e^{2x} = \sum_{k=0}^{\infty} \frac{2^k}{k!} x^k .$$

Substituting in the differential equation, we obtain

$$2c_2 - c_0 + \sum_{k=1}^{\infty} [(k+1)(k+2)c_{k+2} + k c_k - c_k] x^k = \sum_{k=0}^{\infty} \frac{2^k}{k!} x^k$$

If we write few terms of the series on both sides of the equation, we get

$$(2c_2 - c_0) + 6c_3 x + (12c_4 + c_2)x^2 + (20c_5 + 2c_3)x^3 + (30c_6 + 3c_4)x^4$$
$$+ (42c_7 + 4c_5)x^5 = 1 + 2x + 2x^2 + \tfrac{4}{3}x^3 + \tfrac{2}{3}x^4 + \tfrac{5}{15}x^5 + \cdots$$

Equating the coefficients of corresponding powers of x in both sides, we obtain

$$2c_2 - c_0 = 1, \qquad 6c_3 = 2, \qquad 12c_4 + c_2 = 2,$$
$$20c_5 + 2c_3 = \tfrac{4}{3}, \quad 30c_6 + 3c_4 = \tfrac{2}{3}, \quad 42c_7 + 4c_5 = \tfrac{4}{15}$$

Hence, the coefficients in terms of c_0 and c_1 are given by

$$c_2 = \tfrac{1}{2}(1 + c_0), \qquad c_3 = \tfrac{1}{3}, \quad c_4 = \tfrac{1}{24}(3 - c_0),$$
$$c_5 = \tfrac{1}{30}, \quad c_6 = \tfrac{1}{240}(\tfrac{7}{3} + c_0), \qquad c_7 = \tfrac{1}{315}$$

Then the power series solution is

$$y = c_0 + c_1 x + \tfrac{1}{2}(1 + c_0)x^2 + \tfrac{1}{3}x^3 + \tfrac{1}{24}(3 - c_0)x^4 + \tfrac{1}{30}x^5$$
$$+ \tfrac{1}{240}(\tfrac{7}{3} + c_0)x^6 + \tfrac{1}{315}x^7 + \cdots, \quad \text{or}$$

$$y = c_0\left(1 + \tfrac{1}{2}x^2 - \tfrac{1}{24}x^4 + \tfrac{1}{240}x^6 + \cdots\right) + c_1 x$$
$$+ \left(\tfrac{1}{2}x^2 + \tfrac{1}{3}x^3 + \tfrac{3}{24}x^4 + \tfrac{1}{30}x^5 + \tfrac{7}{720}x^6 + \tfrac{1}{315}x^7 + \cdots\right)$$

where c_0 and c_1 are now the two arbitrary constants. The complementary function is displayed in the first line of solution, whereas the particular integral is shown in the second line. Moreover, since the MacLaurin's series expansion for each coefficient functions is convergent for all finite values of x, then the series solution obtained is also convergent for all finite values of x. □

***Example* 5:** Find the power series solution for the initial value problem

$$y''+e^{x}y=1,\ \text{when}\ x=0, y=0\ \text{and}\ y'=1.$$

***Solution*:** The coefficient functions e^x and 1 are analytic at $x=0$, then we assume a series solution of the form

$$y=\sum_{k=0}^{\infty} c_k x^k .$$

Differentiating with respect to x twice, we obtain

$$y'=\sum_{k=1}^{\infty} k c_k x^{k-1}=\sum_{k=0}^{\infty}(k+1)c_{k+1}x^k$$

$$y''=\sum_{k=2}^{\infty} k(k-1)c_k x^{k-2}=\sum_{k=0}^{\infty}(k+1)(k+2)c_{k+2}x^k$$

The function e^{2x} has a power series expansion about $x=0$ given by

$$e^x=\sum_{k=0}^{\infty}\frac{x^k}{k!}.$$

Substituting these values into the differential equation, we get

$$\sum_{k=0}^{\infty}(k+1)(k+2)c_{k+2}x^k+\left(\sum_{k=0}^{\infty}\frac{x^k}{k!}\right)\left(\sum_{k=0}^{\infty}c_k x^k\right)=1.$$

Writing few terms for each summation, we obtain

$$2c_2+6c_2x+12c_4x^2+20c_5x^3+\cdots$$
$$+\left(1+x+\frac{x^2}{2}+\frac{x^3}{6}+\cdots\right)\left(c_0+c_1x+c_2x^2+c_3x^3+\cdots\right)=1,\ \text{or}$$

$$2c_2+6c_3x+12c_4x^2+20c_5x^3+\cdots+c_0+(c_0+c_1)x$$
$$+\left(\tfrac{1}{2}c_0+c_1+c_2\right)x^2+\left(\tfrac{1}{6}c_0+\tfrac{1}{2}c_1+c_2+c_3\right)x^3+\cdots=1$$

Equating the coefficients of the corresponding powers of x in both sides, we get

$$2c_2+c_0=1,\quad 12c_4+\tfrac{1}{2}c_0+c_1+c_2=0$$

$$6c_3+c_0+c_1=0,\quad 20c_5+\tfrac{1}{6}c_0+\tfrac{1}{2}c_1+c_2+c_3=0$$

$$c_2=\tfrac{1}{2}(1-c_0),\quad c_4=-\tfrac{1}{24}(1+2c_1),\quad \text{thus}$$

$$c_3=-\tfrac{1}{6}\left(c_0+c_1\right),\quad c_5=-\tfrac{1}{120}\left(3c_0+2c_1+3\right).$$

The power series solution is

$$y = c_0 + c_1 x + \tfrac{1}{2}(1-c_0)x^2 - \tfrac{1}{6}(c_0+c_1)x^3 - \tfrac{1}{24}(1+2c_1)x^4 - \tfrac{1}{120}(3c_0+2c_1+3)x^5 + \cdots$$

For the initial conditions at $x = 0$ we have $c_0 = 0$ and $c_1 = 1$, then the solution of the initial value problem is

$$y = x + \tfrac{1}{2}x^2 - \tfrac{1}{6}x^3 - \tfrac{1}{8}x^4 - \tfrac{1}{24}x^5 + \cdots. \quad \square$$

***Example* 6**: Find a power series solution for the differential equation:

$$y'' + x^2 y = 0 \text{ about } x = 0.$$

Solution: Assume a solution of the form $y = \sum_{k=0}^{\infty} c_k x^k$.

Then differentiating with respect to x twice, we obtain

$$y' = \sum_{k=1}^{\infty} k c_k x^{k-1} = \sum_{k=0}^{\infty} (k+1)c_{k+1} x^k$$

$$y'' = \sum_{k=2}^{\infty} k(k-1)c_k x^{k-2} = \sum_{k=0}^{\infty} (k+1)(k+2)c_{k+2} x^k$$

Substituting in the differential equation, we obtain

$$\sum_{k=0}^{\infty} (k+1)(k+2)c_{k+2}x^k + x^2 \sum_{k=0}^{\infty} c_k x^k = 0.$$

Equating the coefficients of the general power of x (x^k) to zero, we obtain

$$(k+1)(k+1)c_{k+2} + c_{k-2} = 0, \text{ or } c_{k+2} = -\frac{c_{k-2}}{(k+1)(k+2)}.$$

For $k = 0, 1, 2, \cdots$ in this relation, we obtain

$k = 0: \; c_2 = 0$ $\qquad$ $k = 1: \; c_3 = 0$

$k = 2: \; c_4 = -\dfrac{c_0}{3\cdot 4}$ $\qquad$ $k = 3: \; c_5 = -\dfrac{c_1}{4\cdot 5}$

$k = 4: \; c_6 = -\dfrac{c_2}{5\cdot 6} = 0$ $\qquad$ $k = 5: \; c_7 = -\dfrac{c_3}{6\cdot 7} = 0$

$k = 6: \; c_8 = -\dfrac{c_4}{7\cdot 8} = \dfrac{c_0}{3\cdot 4\cdot 7\cdot 8}$ $\qquad$ $k = 7: \; c_9 = -\dfrac{c_5}{8\cdot 9} = \dfrac{c_1}{4\cdot 5\cdot 8\cdot 9}$

The general solution is

$$y = a\left\{1 - \tfrac{1}{3\cdot 4}x^4 + \tfrac{1}{3\cdot 4\cdot 7\cdot 8}x^8 - \cdots\right\} + b\left\{x - \tfrac{1}{4\cdot 5}x^5 + \tfrac{1}{4\cdot 5\cdot 8\cdot 9}x^9 - \cdots\right\} \square$$

Note: In all the previous examples, we obtained series solutions about $x = 0$. If it is desired to obtain a series solution about any other point, the following example illustrate will this situation.

***Example* 7**: Find a power series solution for the differential equation:

$$y'' + (x-1)^2 y' - 4(x-1)y = 0 \text{ about } x = 1.$$

Solution: First, we let $x = t+1$, the differential Equation becomes

$$\frac{d^2y}{dt^2} + t^2\frac{dy}{dt} - 4ty = 0.$$

Assume a solution of the form $y = \sum_{k=0}^{\infty} c_k t^k$.

Then differentiating with respect to t twice, we obtain

$$y' = \sum_{k=1}^{\infty} k c_k t^{k-1} = \sum_{k=0}^{\infty} (k+1)c_{k+1} t^k$$

$$y'' = \sum_{k=2}^{\infty} k(k-1)c_k t^{k-2} = \sum_{k=0}^{\infty} (k+1)(k+2)c_{k+2} t^k$$

Substituting in the differential equation, we obtain

$$\sum_{k=0}^{\infty} (k+1)(k+2)c_{k+2} t^k + t^2 \sum_{k=0}^{\infty} (k+1)c_{k+1} t^k - 4t \sum_{k=0}^{\infty} c_k t^k = 0$$

Equating the coefficients of the general power of t (t^k) to zero, we obtain

$$(k+1)(k+1)c_{k+2} + (k-1)c_{k-1} - 4c_{k-1} = 0,$$

or $c_{k+2} = -\dfrac{(k-5)c_{k-1}}{(k+1)(k+2)}$, or $c_{k+3} = -\dfrac{(k-4)c_k}{(k+2)(k+3)}$.

For $k = 0, 1, 2, \cdots$, we have

$k = 0:\ c_3 = \dfrac{4}{2\cdot 3}c_0 = \dfrac{2}{3}c_0$ $\qquad$ $k = 1:\ c_4 = \dfrac{3}{3\cdot 4}c_1 = \dfrac{1}{4}c_1$

$k = 2:\ c_5 = \dfrac{2}{4\cdot 5}c_2 = 0$ $\qquad$ $k = 3:\ c_6 = \dfrac{1}{5\cdot 6}c_3 = \dfrac{1}{45}c_0$

$k = 4:\ c_7 = 0$ $\qquad$ $k = 5:\ c_8 = -\dfrac{1}{7\cdot 8}c_5 = 0$

The general solution is

$y = c_0 + c_1 t + \frac{2}{3}c_0 t^3 + \frac{1}{4}c_1 t^4 + \frac{1}{45}c_0 t^6 - \frac{1}{1620}c_0 t^9 + \cdots$, or

$$y = a\left\{1+\tfrac{2}{3}t^3+\tfrac{1}{45}t^6-\tfrac{1}{1620}t^9+\cdots\right\}+b\left\{t+\tfrac{1}{4}t^4\right\}.$$

Back-substituting, we obtain

$$y = a\left\{1+\tfrac{2}{3}(x-1)^3+\tfrac{1}{45}(x-1)^6-\tfrac{1}{1620}(x-1)^9+\cdots\right\}$$

$$+b\left\{(x-1)+\tfrac{1}{4}(x-1)^4\right\} \quad \square$$

Exercise 5.2

a. Find the recurrence relation for the coefficients of the power series solution for the following differential equations about $x = 0$.

1. $y''+xy=0$ **Ans:** $c_{k+2} = -\dfrac{c_k}{(k+1)(k+2)}$

2. $y''-2y'+xy=0$ **Ans:** $c_{k+2} = \dfrac{2c_{k+1}-c_{k-1}}{(k+1)(k+2)},\quad k>0$

3. $y''-x^3y=0$ **Ans:** $c_{k+2} = \dfrac{c_{k-3}}{(k+1)(k+2)},\quad k>2$

4. $y''+(1-x)y'+2xy=0$ **Ans:** $c_{k+2} = -\dfrac{(k+1)c_{k+1}-kc_k-2c_{k-1}}{(k+1)(k+2)},\quad k>0$

5. $y''+xy'+2xy=0$ **Ans:** $c_{k+2} = -\dfrac{k\,c_k+2c_{k-1}}{(k+1)(k+2)},\quad k>0$

6. $y''+y'-x^2y=0$ **Ans:** $c_{k+2} = -\dfrac{(k+1)c_{k+1}-c_{k-2}}{(k+1)(k+2)},\quad k>1$

7. $y''-8xy'=1+2x$ **Ans:** $c_{k+2} = \dfrac{8c_{k-1}}{(k+1)(k+2)},\quad k>1$

b. Find the first four non-zero terms in the power series solution of the following differential equations about $x = 0$.

1. $2y''-4xy'+8x^2y=0$ **Ans:** $y = c_0+c_1x+\tfrac{1}{3}c_1x^3-\tfrac{1}{3}c_0x^4+\cdots$

2. $y''+12y'+x^2y=0$ **Ans:** $y = c_0+c_1x-6c_1x^2+24c_1x^3+\cdots$

3. $y''+2\cos x\cdot y'=x$ **Ans:** $y = c_0+c_1x-c_1x^2+\tfrac{1}{6}(1+4c_1)x^3+\cdots$

4. $y''-2\tan x\cdot y'+y=0$ **Ans:** $y = c_0+c_1x-\tfrac{1}{2}c_0x^2+\tfrac{1}{6}c_1x^3+\cdots$

5. $y''-e^{-3x}y=2x^2$ **Ans:** $y = c_0+c_1x+\tfrac{1}{2}c_0x^2+\tfrac{1}{6}(c_1-3c_0)x^3+\cdots$

c. Find the first five non-zero terms of the solution of the initial value problems.

1. $y''+y'-xy=0$, when $x=0, y=-2$ and $y'=0$

Ans: $y=-2-\frac{1}{3}x^3+\frac{1}{12}x^4-\frac{1}{60}x^5-\frac{1}{120}x^6+\cdots$

2. $y''+2xy'+(x-1)y=0$, when $x=0, y=1$ and $y'=2$

Ans: $y=1+2x+\frac{1}{2}x^2-\frac{1}{2}x^3-\frac{7}{24}x^4+\cdots$

3. $y''-xy=2x$, when $x=1, y=3$ and $y'=0$

Ans: $y=3+\frac{5}{2}(x-1)^2+\frac{5}{6}(x-1)^3+\frac{5}{24}(x-1)^4+\frac{1}{6}(x-1)^5+\cdots$

4. $y''+x^2y=e^x$, when $x=0, y=-2$ and $y'=7$

Ans: $y=-2+7x+\frac{1}{2}x^2+\frac{1}{6}x^3+\frac{1}{24}x^4+\cdots$

d. Solve in series the following Differential Equations near $x=0$:

1. $(1-x^2)y''+2x\,y'-y=0$

Ans: $y=a\left\{1+\frac{1}{2}x^2-\frac{1}{24}x^4-\cdots\right\}+b\left\{x-\frac{1}{6}x^3-\frac{1}{120}x^5-\cdots\right\}$

2. $(x^2+2)y''+x\,y'-(1+x)y=0$

Ans: $y=a\left\{1+\frac{1}{4}x^2+\frac{1}{12}x^3-\frac{3}{96}x^4-\cdots\right\}+b\left\{x+\frac{1}{24}x^4-\cdots\right\}$

3. $(1+x^2)y''+x\,y'-y=0$

Ans: $y=a\left\{1+\frac{1}{2}x^2-\frac{1}{8}x^4+\frac{1}{15}x^6-\cdots\right\}+bx$

4. $(1-x^2)y''-x\,y'+4y=0$

Ans: $y=a\left\{1-2x^2\right\}+b\left\{x-\frac{1}{2}x^3-\frac{1}{8}x^5+\frac{1}{16}x^7+\cdots\right\}$

5. $(2+x^2)y''+x\,y'-xy=1$

Ans: $y=a\left\{1+\frac{1}{4}x^2+\frac{1}{12}x^3-\frac{1}{32}x^4+\cdots\right\}+b\left\{x+\frac{1}{24}x^4+\cdots\right\}$

6. $(1-x^2)y''+2x\,y'+y=0$

Ans: $y=a\left\{1-\frac{1}{2}x^2+\frac{1}{8}x^4+\cdots\right\}+b\left\{x-\frac{1}{2}x^3+\frac{1}{140}x^5+\cdots\right\}$

7. $y''+x\,y'+x^2y=0$

Ans: $y=a\left\{1-\frac{1}{12}x^4+\frac{1}{90}x^6-\cdots\right\}+b\left\{x-\frac{1}{6}x^3-\frac{1}{40}x^5-\cdots\right\}$

8. $(1+x^2)y''+x\,y'-xy=0$

Ans: $y=a\left\{1+\frac{1}{6}x^3-\frac{3}{40}x^5+\cdots\right\}+b\left\{x-\frac{1}{6}x^3+\frac{1}{12}x^4+\frac{3}{45}x^5+\cdots\right\}$

5.4. The Method of Frobenius

In the previous discussion, we dealt with differential equations which are "*well behaved*" at and near x_0. We found that, for linear differential equations, if the coefficient functions are analytic at x_0, then we have no problem finding a power series solution about x_0 by the method of undetermined coefficients or by the use of Taylor's series. Now, what about if one or more of the coefficient functions is not analytic at x_0? Look at this example:

Breakdown example: Find a power series solution for the following differential equation about $x = 0$: $4x^2y'' + y = 0$.

Solution: Assume as before a series solution of the form $y = \sum_{k=0}^{\infty} c_k x^k$.

Differentiating with respect to x twice, we obtain

$$y' = \sum_{k=1}^{\infty} k c_k x^{k-1} = \sum_{k=0}^{\infty} (k+1)c_{k+1} x^k$$

$$y'' = \sum_{k=2}^{\infty} k(k-1)c_k x^{k-2} = \sum_{k=0}^{\infty} (k+1)(k+2)c_{k+2} x^k$$

Substituting in the differential equation, we obtain

$$4\sum_{k=0}^{\infty} k(k-1)c_k x^k + \sum_{k=0}^{\infty} c_k x^k = 0$$

Then equating the coefficient of x^k to zero, we get

$[4k(k-1)+1]c_k = 0$ or $(2k-1)^2 c_k = 0$.

Hence, $c_k = 0$, $k = 0, 1, 2, \cdots$. Thus, we obtain only the trivial solution $y(x) = 0$!!. Were is the power series solution? Something is wrong. In fact, our assumption does not give us the required solution.

In this section, we treat this situation in some details. First, we state some definitions that might help.

Consider the linear second order homogeneous differential equation

$$y'' + P(x)y' + Q(x)y = 0 \tag{6}$$

Definition 1: The point x_0 is an ordinary point of the differential Equation (6) if $P(x)$ and $Q(x)$ are analytic at x_0.

With this definition, if x_0 is an ordinary point of the differential Equation (6), then we can use either the Taylor's series method or the method of undetermined coefficients to find a power series solution about x_0.

Definition 2: If x_0 is not an ordinary point of the differential equation (6), then it is a singular point, look back at the previous example.

Definition 3: The point x_0 is a regular singular point of the differential equation (6) if it is not an ordinary point and if $(x-x_0)P(x)$ and $(x-x_0)^2Q(x)$ are analytic at x_0. Regular singular points are simply called regular points.

Definition 4: If x_0 is neither ordinary nor regular singular point, it is called an irregular singular point.

Example: Find all of the singular points of the following two differential equations and classify each of them as either regular or irregular.

i) $x^2(x-3)^2y''+4x(x^2-x-6)y'+(x^2-x-2)y=0$

ii) $x^{5/2}(x-2)y''-x^{5/2}y'+(x-2)y=0$.

Solution: i) We have $P(x)=\dfrac{4(x+2)}{x(x-3)}$, $Q(x)=\dfrac{(x-2)(x+1)}{x^2(x-3)^2}$.

Then $x=0$ and $x=3$ are singular points for the equation.

For $x=0$, we have $xP(x)=\dfrac{4(x+2)}{x-3}$ and

$x^2Q(x)=\dfrac{(x-2)(x+1)}{(x-3)^2}$, both are analytic at $x=0$. Then $x=0$ is a regular singular point.

For $x=3$, we have

$(x-3)P(x)=\dfrac{4(x+2)}{x}$ and $(x-3)^2Q(x)=\dfrac{(x-2)(x+1)}{x^2}$,

both are analytic at $x=3$. Then $x=3$ is a regular singular point.

ii) We have $P(x)=-\dfrac{1}{x-2}$, $Q(x)=\dfrac{x-2}{x^{5/2}}$, Then $x=0$ and $x=2$ are singular points for the differential equation.

For $x=0$, we have $xP(x)=-\dfrac{x}{x-2}$ and $x^2Q(x)=\dfrac{x-2}{x^{1/2}}$,

the second function is not analytic at $x=0$. Then $x=0$ is an irregular singular point.

For $x=2$, we have

$(x-2)P(x)=-1$ and $(x-2)^2Q(x)=\dfrac{(x-2)^3}{x^{5/2}}$, both are analytic at $x=2$. Then $x=2$ is a regular singular point. □

Solutions about irregular singular points are beyond the scope of this book, so they will not be discussed here. On the other hand, the method of Frobenius deals with solutions of differential equations about regular singular points.

Theorem: If x_0 is a regular point of the differential equation, $y''+P(x)y'+Q(x)y=0$, then there is at least one solution of the form $\sum_{k=0}^{\infty} c_k (x-x_0)^{k+\alpha}$. The interval of convergence of the series solution is not less than the distance from x_0 to the nearest of the other singular points of the equation.

The index α need not be an integer. If we substitute y and its derivatives into the differential equation as we did for the method of undetermined coefficients, we hope to determine the coefficients c's as well as the index α. It is not straightforward as we might expect. The values of α can be either a positive or negative integer, zero or a fraction. Different values of the index α, as we will see, dictate some precautions to be taken.

For the sake of simplification, whenever we wish to obtain solutions about a point other than $x=0$, we first translate the origin to that point and then proceed with the solution procedure. Hence, we concentrate our efforts on solutions valid about $x=0$.

If $x=0$ is a regular singular point of the differential equation

$$y''+P(x)y'+Q(x)y=0 \tag{7}$$

then, we assume a Frobenius series solution of the form

$$y=\sum_{k=0}^{\infty} c_k (x-x_0)^{k+\alpha}, \quad c_0 \neq 0 \tag{8}$$

The index α is obtained, as we shall see, from a quadratic equation in α, called the ***indicial equation***. In general, for each of the two roots of the quadratic equation α_1 and α_2 corresponds a series solution of the differential equation. In some cases, only one series solution is obtained using the Frobenius assumption, and we must find another way to obtain the second solution. Moreover, even is the Frobenius assumption gives us two solutions, these two solutions may not always be linearly independent!

Using the method of Frobenius, the following cases, depending on the values of the index α, will be studied.

Case I: The roots of the indicial equation are distinct and the difference between them is not an integer.

***Example* 1**: Find a series solution for the following differential equation about the origin: $2xy'' + (1+x)y' - 2y = 0$.

Solution: The point $x = 0$ is clearly a regular singular point of the differential equation. Assume a solution of the form

$$y = \sum_{k=0}^{\infty} c_k x^{k+\alpha}, \quad c_0 \neq 0, \text{ then } y' = \sum_{k=0}^{\infty} (k+\alpha) c_k x^{k+\alpha-1}, \text{ and}$$

$$y'' = \sum_{k=0}^{\infty} (k+\alpha)(k+\alpha-1) c_k x^{k+\alpha-2}$$

Substituting in the differential equation, we obtain

$$2x \sum_{k=0}^{\infty} (k+\alpha)(k+\alpha-1) c_k x^{k+\alpha-2} + (1+x) \sum_{k=0}^{\infty} (k+\alpha) c_k x^{k+\alpha-1} - 2\sum_{k=0}^{\infty} c_k x^{k+\alpha} = 0$$

Now, to get the indicial equation, we equate the coefficient of the lowest power of x, ($x^{\alpha-1}$), to zero, we obtain $\alpha(2\alpha-1)c_0 = 0$.

Since $c_0 \neq 0$, the ***indicial equation*** is $\boxed{\alpha(2\alpha-1) = 0}$. The roots are $\alpha = 0$ or $\frac{1}{2}$. The difference between the two roots is not an integer. Then, for each value of α corresponds a solution.

Equating the coefficients of $x^{k+\alpha}$ to zero, we obtain

$(k+\alpha+1)(2k+2\alpha+1)c_{k+1} + (k+\alpha-2)c_k = 0$ or

$$\boxed{c_{k+1} = -\frac{(k+\alpha-2)}{(k+\alpha+1)(2k+2\alpha+1)} c_k, \quad k = 0, 1, 2, \cdots}$$

This is the ***recurrence relation*** for the coefficients.

For $\alpha = 0$: The recurrence relation becomes

$$c_{k+1} = -\frac{(k-2)}{(k+1)(2k+1)} c_k, \quad k = 0, 1, 2, \cdots$$

For various values of k, we have

$k=0: \; c_1 = 2c_0$

$k=1: \; c_2 = \frac{1}{2\cdot 3}c_1 = \frac{1}{3}c_0$

$k=2: \; c_3 = 0$

Then for $k \geq 2: \; c_k = 0$.

The solution corresponding to the root $\alpha = 0$ of the indicial equation is $y_1 = c_0(1+2x+\frac{1}{3}x^2)$, where c_0 is an arbitrary constant.

For $\alpha = 1/2$: The recurrence relation becomes

$$c_{k+1} = -\frac{(2k-3)}{2(2k+3)(k+1)}c_k, \quad k = 0, 1, 2, \cdots$$

For different values of k, we have

$k=0: \; c_1 = -\frac{(-3)}{2\cdot 3}c_0$

$k=1: \; c_2 = -\frac{(-1)}{4\cdot 5}c_1 = \frac{(-3)(-1)}{2\cdot 3\cdot 4\cdot 5}c_0$

$k=2: \; c_3 = -\frac{(1)}{6\cdot}c_2 = -\frac{(-3)(-1)(1)}{2\cdot 3\cdot 4\cdot 5\cdot 6\cdot 7}c_0$

and in general for $k = n-1$, we have

$$c_n = (-1)^n \frac{(-3)(-1)(1)\cdot\cdots\cdot(2n-5)}{(2\cdot 4\cdot 6\cdot\cdots\cdot(2n))[3\cdot 5\cdot 7\cdot\cdots\cdot(2n+1)]}c_0, \quad n<0,$$

and the solution corresponding to the root $\alpha = \frac{1}{2}$ of the indicial equation is

$$y_2 = c_0\sqrt{x}\sum_{n=0}^{\infty}\frac{(-1)^n 3x^n}{2^n n!(2n-3)(2n-1)(2n+1)}.$$

Finally, the general solution is

$$y = A(1+2x+\tfrac{1}{3}x^2) + B\sqrt{x}\sum_{n=0}^{\infty}\frac{(-1)^n 3x^n}{2^n n!(2n-3)(2n-1)(2n+1)}$$

We can use the ratio test to verify that the series appearing in the solution converges for all finite values of x. □

***Example* 2**: Find a series solution for the following differential equation about the origin: $2x^2y''+xy'-(x+1)y=0$.

Solution: Here $P(x)=\dfrac{1}{2x}$ and $Q(x)=-\dfrac{x+1}{2x^2}$. Both $xP(x)$ and $x^2Q(x)$ are analytic at the origin, then the point $x=0$ is a regular singular point of the differential equation. We assume a solution of the form

$$y=\sum_{k=0}^{\infty}c_kx^{k+\alpha},\quad c_0\neq 0\text{, then } y'=\sum_{k=0}^{\infty}(k+\alpha)c_kx^{k+\alpha-1}\text{, and}$$

$$y''=\sum_{k=0}^{\infty}(k+\alpha)(k+\alpha-1)c_kx^{k+\alpha-2}$$

Substituting in the differential equation, we obtain

$$2x^2\sum_{k=0}^{\infty}(k+\alpha)(k+\alpha-1)c_kx^{k+\alpha-2}+x\sum_{k=0}^{\infty}(k+\alpha)c_kx^{k+\alpha-1}$$
$$-(x+1)\sum_{k=0}^{\infty}c_kx^{k+\alpha}=0$$

We equate the coefficient of the lowest power of x, (x^{α}), to zero, to obtain $2\alpha(\alpha-1)c_0+\alpha c_0-c_0=0$. Since $c_0\neq 0$, the ***indicial equation*** is $2\alpha^2-\alpha-1=0$. The roots are $\alpha=1,-\frac{1}{2}$. The difference between the two roots is <u>not an integer</u>. Then, for each value of α corresponds a solution.

Equating the coefficients of $x^{k+\alpha}$ to zero, we obtain

$2(k+\alpha)(k+\alpha-1)c_k+(k+\alpha)c_k-c_{k-1}-c_k=0$ or

$$\boxed{c_k=\frac{c_{k-1}}{(k+\alpha-1)(2k+2\alpha+1)}}.$$

This is the ***recurrence relation*** for the coefficients.

For $\alpha=1$ The recurrence relation becomes $c_k=\dfrac{c_{k-1}}{k(2k+3)}$. Then for various values of k, we have

$k=1$: $c_1=\dfrac{c_0}{1\cdot 5}$ $\qquad$ $k=2$: $c_2=\dfrac{c_1}{2\cdot 7}=\dfrac{c_0}{1\cdot 2\cdot 5\cdot 7}$

$$k=3:\quad c_3=\frac{c_2}{3\cdot 9}=\frac{c_0}{1\cdot 2\cdot 3\cdot 5\cdot 7\cdot 9}\quad\cdots$$

The first solution is

$$y_1=ax\left\{1+\frac{x}{1\cdot 5}+\frac{x^2}{1\cdot 2\cdot 5\cdot 7}+\frac{x^3}{1\cdot 2\cdot 3\cdot 5\cdot 7\cdot 9}+\cdots\right\}.$$

For $\alpha=-1/2$ The recurrence relation becomes $c_k=\dfrac{c_{k-1}}{k(2k-3)}$.

Then for various values of k, we have

$$k=1:\quad c_1=\frac{c_0}{1\cdot(-1)}=-c_0 \qquad\qquad k=2:\quad c_2=\frac{c_1}{2\cdot 1}=-\frac{c_0}{2}$$

$$k=3:\quad c_3=\frac{c_2}{3\cdot 3}=-\frac{c_0}{9}\quad\cdots$$

The second solution is

$$y_2=\frac{b}{\sqrt{x}}\left\{1-x-\frac{x^2}{2}-\frac{x^3}{9}-\cdots\right\}.$$

Finally, the general solution will be

$$y=ax\left\{1+\frac{x}{1!5}+\frac{x^2}{2!5\cdot 7}+\frac{x^3}{3!5\cdot 7\cdot 9}+\cdots\right\}$$
$$+\frac{b}{\sqrt{x}}\left\{1-x-\frac{x^2}{2}-\frac{x^3}{9}-\cdots\right\}\square$$

***Example* 3**: Find a series solution for the following differential equation about the origin: $2x(1-x)y''+(1-x)y'+3y=0$.

Solution: Here $P(x)=\dfrac{1}{2x}$ and $Q(x)=\dfrac{3}{2x(1-x)}$. Both $xP(x)$ and $x^2Q(x)$ are analytic at $x=0$, then the point $x=0$ is a regular singular point of the differential equation. We assume a solution of the form

$$y=\sum_{k=0}^{\infty}c_kx^{k+\alpha},\quad c_0\neq 0\text{, then } y'=\sum_{k=0}^{\infty}(k+\alpha)c_kx^{k+\alpha-1}\text{, and}$$

$$y''=\sum_{k=0}^{\infty}(k+\alpha)(k+\alpha-1)c_kx^{k+\alpha-2}$$

Substituting in the differential equation, we obtain

$$(2x - 2x^2)\sum_{k=0}^{\infty}(k+\alpha)(k+\alpha-1)c_k x^{k+\alpha-2}$$
$$+(1-x)\sum_{k=0}^{\infty}(k+\alpha)c_k x^{k+\alpha-1} + 3\sum_{k=0}^{\infty} c_k x^{k+\alpha} = 0$$

We equate the coefficient of the lowest power of x, ($x^{\alpha-1}$), to zero, to obtain $2\alpha(\alpha-1)c_0 + \alpha c_0 = 0$. Since $c_0 \neq 0$, the ***indicial equation*** is $2\alpha^2 - \alpha = 0$. The roots are $\alpha = 0, \frac{1}{2}$. The difference between the two roots is not an integer. Then, for each value of α corresponds a solution.

Equating the coefficients of $x^{k+\alpha}$ to zero, we obtain

$$2(k+\alpha+1)(k+\alpha)c_{k+1} - 2(k+\alpha)(k+\alpha-1)c_k$$
$$+(k+\alpha+1)c_{k+1} - (k+\alpha)c_k + 3c_k = 0$$

or

$$\boxed{c_{k+1} = \frac{2k+2\alpha-3}{2k+2\alpha+1}c_k}.$$

This is the ***recurrence relation*** for the coefficients.

For $\alpha = 0$ The recurrence relation becomes $c_{k+1} = \dfrac{(2k-3)c_k}{2k+1}$.

Then for various values of k, we have

$k=0:\ c_1 = (-3)c_0$ $\qquad$ $k=1:\ c_2 = \dfrac{-c_1}{3} = \dfrac{(-3)(-1)c_0}{1\cdot 3}$

$k=2:\ c_3 = \dfrac{c_2}{5} = \dfrac{(-3)(-1)(1)\ldots c_0}{1\cdot 3\cdot 5}$ . . ., in general

$$c_n = \frac{(-3)(-1)\cdot 1\cdot 3\cdot 5\cdot\dots\cdot(2n-5)c_0}{1\cdot 3\cdot 5\cdot\dots\cdot(2n-1)} = \frac{3c_0}{(2n-1)(2n-3)}, \quad n \geq 1.$$

The first solution is

$$y_1 = a\left\{1 + \sum_{n=1}^{\infty}\frac{3x^n}{(2n-1)(2n-3)}\right\}.$$

For $\alpha = 1/2$ The recurrence relation becomes $c_{k+1} = \dfrac{(k-1)c_k}{k+1}$.

Then for various values of k, we have

$k=0:\ c_1 = -c_0$ $\qquad$ $k=1:\ c_2 = 0,\ c_3 = c_4 = c_5 = \cdots = 0.$

The second solution is

$$y_2 = b\sqrt{x}\,\{1-x\}.$$

Finally, the general solution will be

$$y = a\left\{1+\sum_{n=1}^{\infty}\frac{3x^n}{(2n-1)(2n-3)}\right\}+b\sqrt{x}\,\{1-x\}.$$ □

Exercise 5.3

Obtain the series solution of the following differential equations about $x=0$

1. $2x(x-1)y''+3(x-1)y'-y=0$

$$\textbf{Ans: } y = A\left(1-\sum_{n=1}^{\infty}\frac{x^n}{4n^2-1}\right)+\left(x^{-1/2}-x^{1/2}\right)$$

2. $2x^2(x+1)y''+x(7x-1)y'+y=0$ **Ans:**

$$y = A\sum_{n=0}^{\infty}(-1)^n(2n+3)(2n+5)x^{n+1}+B\sum_{n=0}^{\infty}(-1)^n(n+1)(n+2)x^{n+1/2}$$

3. $3xy''+(2-x)y'-2y=0$

$$\textbf{Ans: } y = A\sum_{n=0}^{\infty}\frac{(3n+4)}{4\cdot 3^n\, n!}x^{n+1/3}+B\left(1+\sum_{n=1}^{\infty}\frac{(n+1)x^n}{2\cdot 5\cdot 8\cdot\cdots\cdot(3n-1)}\right)$$

4. $2x^2y''-x(2x+1)y'+(1-5x)y=0$ **Ans:**

$$y = A\sum_{n=0}^{\infty}\frac{(2n+3)(2n+5)}{n!}x^{n+1}+B\left(x^{1/2}+\sum_{n=1}^{\infty}\frac{2^{n-1}(n+1)(n+2)}{1\cdot 3\cdot 5\cdot\cdots\cdot(2n-1)}x^{n+1/2}\right)$$

5. $2xy''+(1-2x^2)y'-4xy=0$

$$\textbf{Ans: } y = A\,x^{1/2}e^{x^2/2}+B\left(1+\sum_{n=1}^{\infty}\frac{2^n x^{2n}}{3\cdot 7\cdot 11\cdot\cdots\cdot(4n-1)}\right)$$

6. $2xy''+(1+2x)y'-5y=0$

$$\textbf{Ans: } y = A\,x^{1/2}\left(1+\tfrac{4}{3}x+\tfrac{4}{15}x^2\right)+B\sum_{n=0}\frac{15(-1)^{n+1}x^n}{n!(2n-5)(2n-3)(2n-1)}$$

7. $9x^2y''+3x(x+3)y'-(1+4x)y=0$

$$\textbf{Ans: } y=A\,x^{1/3}(1+\tfrac{1}{5}x)+B\sum_{n=0}^{\infty}\frac{10(-1)^n x^{n-1/3}}{3^n\,n!(3n-5)(3n-2)}$$

8. $2x^2y''+5xy'-2y=0$ **Ans:** $y=A\,x^{1/2}+B\,x^{-2}$

9. $4x\,y''+2y'+y=0$

Ans:
$$y=a\left\{1-\tfrac{1}{2!}x+\tfrac{1}{4!}x^2-\tfrac{1}{6!}x^3+\cdots\right\}+b\sqrt{x}\left\{1-\tfrac{1}{3!}x+\tfrac{1}{5!}x^2+\cdots\right\}$$
$$=a\cos\sqrt{x}+b\sin\sqrt{x}$$

10. $8x^2y''-2x\,y'+y=0$ **Ans:** $y=a\sqrt{x}+b\sqrt[4]{x}$

11. $9x(1-x)y''-12y'+4y=0$

$$\textbf{Ans: } y=a\left\{1+\tfrac{1}{3}x+\tfrac{1\cdot4}{3\cdot6}x^2+\cdots\right\}+b\,x^{7/3}\left\{1+\tfrac{8}{10}x+\tfrac{8\cdot11}{10\cdot13}x^2+\cdots\right\}$$

12. $2x^2y''-x\,y'+(1-x^2)y=0$

$$\textbf{Ans: } y=ax\left\{1+\frac{x^2}{2\cdot5}+\frac{x^4}{2\cdot4\cdot5\cdot9}+\cdots\right\}+b\sqrt{x}\left\{1+\frac{x^2}{2\cdot3}+\frac{x^4}{2\cdot3\cdot4\cdot7}+\cdots\right\}$$

13. $4x\,y''+2(1-x)y'-y=0$

$$\textbf{Ans: } y=a\left\{1+\frac{x}{2\cdot1!}+\frac{x^2}{2^2\cdot2!}+\frac{x^3}{2^3\cdot3!}+\cdots\right\}+b\sqrt{x}\left\{1+\frac{x}{1\cdot3}+\frac{x^2}{1\cdot3\cdot5}+\frac{x^3}{1\cdot3\cdot5\cdot7}+\cdots\right\}$$

14. $3x\,y''+(1-x)y'-y=0$

$$\textbf{Ans: } y=a\left\{1+x+\frac{x^2}{4}+\frac{x^3}{4\cdot7}+\cdots\right\}+b\,x^{2/3}\left\{1+\frac{x}{3}+\frac{x^2}{3\cdot6}+\frac{x^3}{3\cdot6\cdot9}+\cdots\right\}$$

15. $2x(1-x)y''+(1-x)y'+3y=0$

$$\textbf{Ans: } y=a\left\{1-3x+\frac{3x^2}{1\cdot3}+\frac{3x^3}{3\cdot5}+\frac{3x^4}{5\cdot7}\cdots\right\}+b(1-x)\sqrt{x}$$

Case II: The two roots of the indicial equation are equal

In this case, $\alpha_1 = \alpha_2 = a$, and we obtain only one Frobenius series solution. Suppose that this solution in term of the double root α is $y_1 = g(x, \alpha)$, then if we substitute this function in the differential equation we get

$$\frac{\partial^2 g}{\partial x^2} + P(x)\frac{\partial g}{\partial x} + Q(x)g = F(x, \alpha). \tag{9}$$

We have used partial derivatives since g is a function of two variables x and α. You may notice the function $F(x, \alpha)$ to the right of Equation (9). This function vanishes when $\alpha = a$, since $g(x, a)$ is a solution of the differential equation. In fact, because $\alpha = a$ is a double root, we will also have $\left.\frac{\partial}{\partial \alpha}F(x, \alpha)\right|_{\alpha=a} = 0$.

Now, differentiating Equation (9) with respect to α, we obtain

$$\frac{\partial}{\partial \alpha}\left(\frac{\partial^2 g}{\partial x^2}\right) + P(x)\frac{\partial}{\partial \alpha}\left(\frac{\partial g}{\partial x}\right) + Q(x)\frac{\partial}{\partial \alpha}(g) = \frac{\partial}{\partial \alpha}F(x, \alpha)$$

Interchanging the order of differentiation, we obtain

$$\frac{\partial^2}{\partial x^2}\left(\frac{\partial g}{\partial \alpha}\right) + P(x)\frac{\partial^2}{\partial x^2}\left(\frac{\partial g}{\partial \alpha}\right) + Q(x)\left(\frac{\partial g}{\partial \alpha}\right) = \frac{\partial}{\partial \alpha}F(x, \alpha) \tag{10}$$

Clearly, $\frac{\partial g}{\partial \alpha}$ satisfies the differential equation when $\alpha = a$, hence

$$y_1 = g(x, a) \quad \text{and} \quad y_2 = \left.\frac{\partial}{\partial \alpha}g(x, \alpha)\right|_{\alpha=a}$$

are the two solutions of the equation. They are, in fact, linearly independent.

Another approach is as follows. Since, in the case of a double root, Frobenius method produces only one series solution, say y_1, given by

$$y_1 = x^a \sum_{k=0}^{\infty} c_k x^k,$$

We can use the method of variation of parameters to obtain the second linearly independent solution by assuming a solution of the form

$$y(x) = y_1 \cdot v(x).$$

In this case, we will end up with

$$v(x) = \ln x + \sum_{k=1}^{\infty} \beta_k x^{k+a}.$$

***Example* 1**: Find the series solution for the following equation about $x=0$:

$$(x^2-x)y''+(2x+1)y'-\frac{1}{x}y=0.$$

Solution: The coefficient functions are

$$P(x)=\frac{2x+1}{x(x-1)} \text{ and } Q(x)=-\frac{1}{x^2(x-1)}.$$

Then $xP(x)$ and $x^2Q(x)$ are analytic at $x=0$. Then $x=0$ is a regular point. Assume a Frobenius series solution of the form

$$y=\sum_{k=0}^{\infty}c_k x^{k+\alpha},\quad c_0\neq 0\text{, then } y'=\sum_{k=0}^{\infty}(k+\alpha)c_k x^{k+\alpha-1}\text{, and}$$

$$y''=\sum_{k=0}^{\infty}(k+\alpha)(k+\alpha-1)c_k x^{k+\alpha-2}$$

Substituting for y and its derivatives into the differential equation, we get

$$(x^2-x)\sum_{k=0}^{\infty}(k+\alpha)(k+\alpha-1)c_k x^{k+\alpha-2}$$
$$+(2x+1)\sum_{k=0}^{\infty}(k+\alpha)c_k x^{k+\alpha-1}-\frac{1}{x}\sum_{k=0}^{\infty}c_k x^{k+\alpha}=0$$

To get the indicial equation, we equate the coefficient of the lowest power of x, which is $x^{\alpha-1}$, to zero, we obtain

$(\alpha-1)^2c_0=0$, and since $c_0\neq 0$, we have $(\alpha-1)^2=0$. Then

$\alpha=1,\ 1$. We have a double root. Then, one solution is $g(x,1)$ corresponding to $\alpha=1$, and the other is $y_2=\left.\frac{\partial}{\partial\alpha}g(x,\alpha)\right|_{\alpha=1}$.

Now, equating the coefficients of $x^{k+\alpha}$ to zero, we obtain

$$[(k+\alpha)(k+\alpha-1)+2(k+\alpha)]c_k$$
$$+[-(k+\alpha+1)(k+\alpha)+(k+\alpha+1)-1]c_{k+1}=0.$$

Rearranging, we get the recurrence relation

$$c_{k+1} = \frac{k+\alpha+1}{k+\alpha} c_k, \; k \geq 0, \; \alpha = 1 \text{ and } k+\alpha \neq 0$$

For various values of k, we have

$k = 0$: $\quad c_1 = \dfrac{\alpha+1}{\alpha} c_0$

$k = 1$: $\quad c_2 = \dfrac{\alpha+2}{\alpha} c_0$

$k = 2$: $\quad c_3 = \dfrac{\alpha+3}{\alpha} c_0$

And for $k = n-1$: $\quad c_n = \dfrac{\alpha+n}{\alpha} c_0$

Then, the solution in terms of α is $y = c_0 \sum_{n=0}^{\infty} \dfrac{\alpha+n}{\alpha} x^{n+\alpha}$.

Differentiating with respect to α, we obtain

$$\frac{\partial y}{\partial \alpha} = c_0 \ln x \sum_{n=0}^{\infty} \frac{n+\alpha}{\alpha} x^{n+\alpha} - c_0 \sum_{n=0}^{\infty} \frac{n}{\alpha^2} x^{n+\alpha}.$$

Then the two solutions are

$$y_1 = y(x,1) = c_0 \sum_{n=0}^{\infty} (n+1) x^{n+1}, \text{ and}$$

$$y_2 = \left.\frac{\partial y}{\partial \alpha}\right|_{\alpha=1} = c_0 \left(\ln x \sum_{n=0}^{\infty} (n+1) x^{n+1} - \sum_{n=0}^{\infty} n x^{n+1} \right).$$

The general solution is

$$y = A \sum_{n=0}^{\infty} (n+1) x^{n+1} + B \left(\ln x \sum_{n=0}^{\infty} (n+1) x^{n+1} - \sum_{n=0}^{\infty} n x^{n+1} \right) \quad \square$$

***Example* 2:** Find the series solution for the following equation about $x = 0$:

$$x y'' + y' + x y = 0.$$

***Solution*:** The coefficient functions are $P(x) = 1/x$ and $Q(x) = 1$. Then $xP(x)$ and $x^2 Q(x)$ are analytic at $x = 0$. Then $x = 0$ is a regular point. Assume a Frobenius series solution of the form

$y=\sum_{k=0}^{\infty} c_k x^{k+\alpha}$, $c_0 \neq 0$, then $y'=\sum_{k=0}^{\infty}(k+\alpha)c_k x^{k+\alpha-1}$, and

$$y''=\sum_{k=0}^{\infty}(k+\alpha)(k+\alpha-1)c_k x^{k+\alpha-2}$$

Substituting for y and its derivatives into the differential equation, we get

$$x\sum_{k=0}^{\infty}(k+\alpha)(k+\alpha-1)c_k x^{k+\alpha-2}$$
$$+\sum_{k=0}^{\infty}(k+\alpha)c_k x^{k+\alpha-1}+x\sum_{k=0}^{\infty} c_k x^{k+\alpha}=0$$

Equating the coefficient of the lowest power of x, ($x^{\alpha-1}$), to zero, we obtain $\alpha(\alpha-1)c_0+\alpha c_0=0$. Since $c_0 \neq 0$, the ***indicial equation*** is $\alpha^2=0$. The roots are $\boxed{\alpha = 0,\ 0}$. We have a double root.

Equating the coefficient of the next lowest power of x, (x^{α}), to zero, we obtain $\alpha(\alpha+1)c_1+(\alpha+1)c_1=0$, or $(\alpha+1)^2 c_1=0$. This implies that $\boxed{c_1=0}$.

Equating the coefficients of $x^{k+\alpha}$ to zero, we obtain

$(k+\alpha+1)(k+\alpha)c_{k+1}+(k+\alpha+1)c_k+c_{k-1}=0$ or

$$\boxed{c_{k+1}=-\frac{c_{k-1}}{(k+\alpha+1)^2}}.$$

This is the ***recurrence relation*** for the coefficients.

For various values of k, we have

$k=1$: $c_2=-\dfrac{c_0}{(\alpha+2)^2}$ $\qquad$ $k=2$: $c_3=-\dfrac{c_1}{(\alpha+3)^2}=0$

$k=3$: $c_4=\dfrac{c_0}{(\alpha+2)^2(\alpha+4)^2}$ $\qquad$ $\boxed{c_{2n+1}=0,\ n=0,1,2,3,\cdots}$

$k=5$: $c_6=-\dfrac{c_0}{(\alpha+2)^2(\alpha+4)^2(\alpha+6)^2}$, and

$$\boxed{c_{2n}=\frac{(-1)^n c_0}{(\alpha+2)^2(\alpha+4)^2\cdots(\alpha+2n)^2},\quad n=1,2,3,\cdots}.$$

The solution takes the form

$$\boxed{y = x^{\alpha}\left\{1-\sum_{n=1}^{\infty}\frac{(-1)^n x^{2n}}{(\alpha+2)^2(\alpha+4)^2\cdots(\alpha+2n)^2}\right\}}$$

The first solution is obtained by letting $\alpha = 0$

$$y_1 = \left\{1-\sum_{n=1}^{\infty}\frac{(-1)^n x^{2n}}{2^2.4^2\cdots\cdots(2n)^2}\right\}.$$

Now, we know that the second solution is $y_2 = \left.\frac{\partial y}{\partial\alpha}\right|_{\alpha=0}$, then differentiation the expression of y with respect to α, we obtain

$$\frac{\partial y}{\partial\alpha} = x^{\alpha}\ln x\left\{1-\sum_{n=1}^{\infty}\frac{(-1)^n x^{2n}}{(\alpha+2)^2(\alpha+4)^2\cdots(\alpha+2n)^2}\right\}$$
$$-x^{\alpha}\sum_{n=1}^{\infty}(-1)^n x^{2n}\frac{d}{d\alpha}\left[\frac{1}{(\alpha+2)^2(\alpha+4)^2\cdots(\alpha+2n)^2}\right]$$

To obtain $\frac{d}{d\alpha}\left[\frac{1}{(\alpha+2)^2(\alpha+4)^2\cdots(\alpha+2n)^2}\right] = \frac{dw}{d\alpha}$, we use logarithmic differentiation to get

$$\frac{dw}{d\alpha} = \frac{-2}{(\alpha+2)^2(\alpha+4)^2\cdots(\alpha+2n)^2}\left[\frac{1}{\alpha+2}+\frac{1}{\alpha+4}+\ldots+\frac{1}{\alpha+2n}\right].$$

Finally,

$$y_2 = \left.\frac{\partial y}{\partial\alpha}\right|_{\alpha=0} = \ln x\left\{1-\sum_{n=1}^{\infty}\frac{(-1)^n x^{2n}}{2^2 4^2\ldots(2n)^2}\right\}+\sum_{n=1}^{\infty}(-1)^n x^{2n}\left.\frac{dw}{d\alpha}\right|_{\alpha=0}$$

$$y_2 = y_1\ln x + 2\sum_{n=1}^{\infty}\frac{(-1)^n x^{2n}}{2^2.4^2\cdots\cdots(2n)^2}\left[\tfrac{1}{2}+\tfrac{1}{4}+\cdots+\tfrac{1}{2n}\right].$$

The general solution is

$$y = a\left\{1-\sum_{n=1}^{\infty}\frac{(-1)^n x^{2n}}{2^2.4^2\cdots\cdots(2n)^2}\right\}$$
$$+b\,y_1\ln x + 2\sum_{n=1}^{\infty}\frac{(-1)^n x^{2n}}{2^2.4^2\cdots\cdots(2n)^2}\left[\tfrac{1}{2}+\tfrac{1}{4}+\cdots+\tfrac{1}{2n}\right]\square$$

The last two examples reveal that the second solution takes the form

$$\boxed{y_2 = y_1 \ln x + \sum_{n=1}^{\infty} \beta_n x^{n+\alpha}}.$$

In fact, this is always the case when the indicial equation has a double root. We state here a theorem for this particular case. The proof is omitted.

Theorem: If $x = 0$ is a regular singular point of the differential equation: $y'' + P(x)y' + Q(x)y = 0$, and if the roots of the corresponding indicial equation are equal ($\alpha_1 = \alpha_2$), then there is a Frobenius series solution

$$y_1 = \sum_{n=0}^{\infty} c_n x^{n+\alpha_1}, \quad c_0 \neq 0 \tag{11}$$

Also, the second linearly independent solution is given by

$$y_2 = y_1 \ln x + \sum_{n=1}^{\infty} \beta_n x^{n+\alpha_1}. \tag{12}$$

The coefficients β_n are obtained by substituting y_2 and its derivatives into the differential equation. By doing so, we obtain a recurrence relation for β_n. The following example illustrates the use of this theorem.

***Example* 3**: Find the series solution for the following equation about $x = 0$: $xy'' + (x+1)y' + 2y = 0$

Solution: It is routine to check that $x = 0$ is a regular singular point of this differential equation. Assume a Frobenius series solution of the form

$$y = \sum_{k=0}^{\infty} c_k x^{k+\alpha}, \quad c_0 \neq 0, \text{ then } y' = \sum_{k=0}^{\infty} (k+\alpha)c_k x^{k+\alpha-1}, \text{ and}$$

$$y'' = \sum_{k=0}^{\infty} (k+\alpha)(k+\alpha-1)c_k x^{k+\alpha-2}$$

Substituting for y and its derivatives in the differential equation, we get

$$x\sum_{k=0}^{\infty} (k+\alpha)(k+\alpha-1)c_k x^{k+\alpha-2}$$
$$+ (x+1)\sum_{k=0}^{\infty} (k+\alpha)c_k x^{k+\alpha-1} + 2\sum_{k=0}^{\infty} c_k x^{k+\alpha} = 0$$

To get the indicial equation, we equate the coefficient of the lowest power of *x*, which is $x^{\alpha-1}$, to zero, we obtain $\alpha^2 c_0 = 0$, and since $c_0 \neq 0$, we have $\alpha^2 = 0$. Then $\alpha = 0,\ 0$. We have a double root. Then, one solution is $g(x,0)$ corresponding to $\alpha = 0$, and the other is

$$y_2 = \frac{\partial}{\partial \alpha} g(x,\alpha)\bigg|_{\alpha=0}.$$

Now, equating the coefficients of $x^{k+\alpha}$ to zero, we obtain

$$(k+\alpha-1)^2 c_{k+1} + (k+\alpha+2)c_k = 0.$$

Rearranging, we get the recurrence relation

$$c_{k+1} = -\frac{k+\alpha+2}{(k+\alpha+1)^2} c_k,\ k \geq 0.$$

<u>For $\alpha = 0$</u>, we have $c_{k+1} = -\dfrac{k+2}{(k+1)^2} c_k,\ k \geq 0.$

For various values of *k*, we have

$k = 0$: $\quad c_1 = -\dfrac{2}{1^2} c_0$

$k = 1$: $\quad c_2 = -\dfrac{3}{2^2} c_1 = \dfrac{2 \cdot 3}{1^2 \cdot 2^2} c_0$

$k = 2$: $\quad c_3 = -\dfrac{4}{3^2} c_2 = -\dfrac{2 \cdot 3 \cdot 4}{1^2 \cdot 2^2 \cdot 3^2} c_0$

And for $k = n-1$: $\quad c_n = (-1)^n \dfrac{(n+1)}{n!} c_0$

The first solution is $y_1 = c_0 \displaystyle\sum_{n=0}^{\infty} (-1)^n \frac{(n+1)}{n!} x^n$.

Now, according to the previous theorem, the second solution will be of the form

$$y_2 = y_1 \ln x + \sum_{n=1}^{\infty} \beta_n x^n.$$

Differentiating twice with respect to x, we obtain

$$y_2' = \frac{1}{x} y_1 + y_1' \ln x + \sum_{n=1}^{\infty} n\,\beta_n x^{n-1} \quad \text{and}$$

$$y_2'' = -\frac{1}{x^2} y_1 + \frac{2}{x} y_1' + y_1'' \ln x + \sum_{n=2}^{\infty} n(n-1)\beta_n x^{n-2}.$$

Substituting these values into the differential equation, we obtain

$$-\frac{1}{x} y_1 + 2y_1' + x y_1'' \ln x + \sum_{n=2}^{\infty} n(n-1)\beta_n x^{n-1} + y_1 + x y_1' \ln x$$
$$+\sum_{n=1}^{\infty} n\,\beta_n x^n + \frac{1}{x} y_1 + y_1' \ln x + \sum_{n=1}^{\infty} n\,\beta_n x^{n-1} + 2y_1 \ln x + 2\sum_{n=1}^{\infty} \beta_n x^n$$

We can see that lnx is multiplied by $[x y_1'' + (x+1) y_1' + 2y_1]$ and since y_1 is a solution of the differential equation then

$$x y_1'' + (x+1) y_1' + 2y_1 = 0,$$

and the term containing $\ln x$ vanishes. Rearranging and simplifying, we get

$$2y_1' + y_1 + \sum_{n=1}^{\infty} [n(n+1)\beta_{n+1} + (n+2)\beta_n] x^n + \sum_{n=1}^{\infty} n\,\beta_n x^{n-1} = 0$$

Now, substituting for y_1 (let $c_0 = 1$ for convenience), we obtain

$$2\sum_{n=1}^{\infty} (-1)^n \frac{n(n+1)}{n!} x^{n-1} + \sum_{n=0}^{\infty} (-1)^n \frac{(n+1)}{n!} x^n$$
$$+\sum_{n=1}^{\infty} [n(n+1)\beta_{n+1} + (n+2)\beta_n] x^n + \sum_{n=1}^{\infty} n\,\beta_n x^{n-1} = 0$$

Equating the coefficient of x^0 to zero, we get $\beta_1 = 3$.

Equating the coefficient of x^n to zero, we get the recurrence relation

$$\beta_{n+1} = (-1)^n \frac{n+3}{(n+1)^2 n!} - \frac{n+2}{(n+1)^2} \beta_n, \; n \ge 1.$$

From which

$$\beta_2 = -\tfrac{13}{4},\ \beta_3 = \tfrac{31}{18},\ \beta_4 = -\tfrac{163}{228}, \cdots,$$

and the second solution is

$$y_2 = y_1 \ln x + \left(3x - \tfrac{13}{4}x^2 + \tfrac{31}{18}x^3 - \tfrac{163}{228}x^4 + \cdots\right)$$

and the general solution is

$$y = A\sum_{n=0}^{\infty}(-1)^n \frac{(n+1)}{n!}x^n + B\left(\begin{array}{l}\ln x \displaystyle\sum_{n=0}^{\infty}(-1)^n \frac{(n+1)}{n!}x^n \\ +\left(3x - \tfrac{13}{4}x^2 + \tfrac{31}{18}x^3 + \cdots\right)\end{array}\right) \quad \square$$

There is another approach to obtain the second solution. The solution in terms of α is given by

$$y = x^{\alpha}\left\{1 - \frac{\alpha+2}{(\alpha+1)^2}x + \frac{\alpha+3}{(\alpha+1)^2(\alpha+2)}x^2 - \frac{\alpha+4}{(\alpha+1)^2(\alpha+2)(\alpha+3)}x^3 + \cdots\right\}.$$

Differentiating this expression with respect to α, we obtain

$$\frac{\partial y}{\partial \alpha} = x^{\alpha}\ln x\left\{1 - \frac{\alpha+2}{(\alpha+1)^2}x + \frac{\alpha+3}{(\alpha+1)^2(\alpha+2)}x^2 - \frac{\alpha+4}{(\alpha+1)^2(\alpha+2)(\alpha+3)}x^3 + \cdots\right\}$$

$$+x^{\alpha}\left\{-\frac{\alpha+2}{(\alpha+1)^2}\left[\frac{1}{\alpha+2} - \frac{2}{\alpha+1}\right]x + \frac{\alpha+3}{(\alpha+1)^2(\alpha+2)}\left[\frac{1}{\alpha+3} - \frac{2}{\alpha+1} - \frac{1}{\alpha+2}\right]x^2 + \cdots\right\}$$

Letting $\alpha = 0$ in this last expression, we obtain the second solution as

$$y_2(x) = \ln x\left\{\sum_{n=0}^{\infty}(-1)^n \frac{n+1}{n!}x^n\right\}$$

$$+\left\{3x - \frac{3}{2!}\left[2 + \frac{1}{2} - \frac{1}{3}\right]x^2 + \frac{4}{3!}\left[2 + \frac{1}{2} + \frac{1}{3} - \frac{1}{4}\right]x^3 + \cdots\right\}.$$

The general solution will be

$$y = A\sum_{n=0}^{\infty}(-1)^n \frac{(n+1)}{n!}x^n + B\left(\begin{array}{l}\ln x \displaystyle\sum_{n=0}^{\infty}(-1)^n \frac{(n+1)}{n!}x^n \\ +\left(3x - \tfrac{13}{4}x^2 + \tfrac{31}{18}x^3 + \cdots\right)\end{array}\right) \quad \square$$

Exercise 5.4

Obtain the two linearly independent solutions of the following equation using the method of Frobenius:

1. $x(x-1)y''+(3x-1)y'+y=0$ **Ans:** $y_1=\sum_{n=0}^{\infty} x^n,\ y_2=y_1\ln x$

2. $xy''-y'-xy=0$

Ans: $y_1=1+\sum_{n=1}^{\infty}\frac{x^{2n}}{2^2\cdot 4^2\cdot 6^2\cdots\cdots(2n)^2},\ y_2=y_1\ln x-\frac{x^2}{4}-\frac{3x^4}{8\cdot 16}-\cdots$

3. $x^2y''+3xy'+(1-2x)y=0$

Ans: $y_1=\frac{1}{x}+\sum_{n=1}^{\infty}\frac{2^n x^{n-1}}{(n!)^2},\ y_2=y_1\ln x-\sum_{n=1}^{\infty}\frac{2^{n+1}\left(1+\frac{1}{2}+\frac{1}{3}+\cdots+\frac{1}{n}\right)}{(n!)^2}x^{n-1}$

4. $x^2y''+x(x-1)y'+(1-x)y=0$

Ans: $y_1=x,\ y_2=y_1\ln x+\sum_{n=1}^{\infty}\frac{(-1)^n x^{n+1}}{n\cdot n!}$

5. $x(1+x)y''+(1+5x)y'+3y=0$

Ans: $y_1=1+\frac{1}{2}\sum_{n=1}^{\infty}(-1)^n(n+1)(n+2)x^n$

$y_2=y_1\ln x-\frac{3}{2}(y_1-1)+\frac{1}{2}\sum_{n=1}^{\infty}(-1)^n(2n+3)x^n$

6. $x(x-2)y''+2(x-1)y'-2y=0$

Ans: $y_1=1-x,\ y_2=y_1\ln x+\frac{5}{2}x-\sum_{n=2}^{\infty}\frac{(n+1)x^n}{2^n n(n-1)}$

7. Show that the series solution of the equation $xy''+y'+y=0$ is of the form $y=(A+B\ln x)\sum_{n=0}^{\infty}c_nx^n+B\sum_{n=0}^{\infty}\beta_nx^n$, and find the values of the coefficients c_n and β_n for $n=0,1,2,3$.

Case III: The roots of the indicial equation are distinct differ by an integer

This case is best described by the following theorem.

Theorem*: If $x = 0$ is a regular singular point of the differential equation: $y'' + P(x)y' + Q(x)y = 0$, and if α_1 and α_2 are the two roots of the indicial equation, and if $(\alpha_1 - \alpha_2)$ is a positive integer, then there is a Frobenius series solution of the form

$$y_1 = \sum_{n=0}^{\infty} c_n x^{n+\alpha_1}, \; c_0 \neq 0 \tag{13}$$

Also, the second linearly independent solution is given by

$$y_2 = m y_1 \ln x + \sum_{n=0}^{\infty} \beta_n x^{n+\alpha_2}. \tag{14}$$

The constants m and β_n is obtained by substituting expression (14) into the differential equation and equating the coefficients of various powers of x to zero.

The number m may or may not be zero. If $m = 0$, then the solution y_2 will not contain a logarithmic term and we have the ***Nonlogarithmic case*** (see ***Example*** 1 below). On the other hand, if $m \neq 0$, the solution y_2 will contain a logarithmic term, and we have the ***Logarithmic case*** (see ***Example*** 2 below).

The solution procedure is then:

Step 1: Obtain the first solution y_1 by assuming a Frobenius series solution of the form (13) then evaluating the coefficients c_n.

Step 2: Substitute y_2 from expression (14) into the differential equation to obtain the constant m and the coefficients β_n.

Step 3: The general solution is $y = Ay_1 + By_2$.

Example **1**: ***Nonlogarithmic case***: Find the series solution for the following equation about $x = 0$: $xy'' - (4 + x)y' + 2y = 0$

Solution: The point $x = 0$ is a regular singular point of this differential equation. Assume a Frobenius series solution of the form

$$y = \sum_{k=0}^{\infty} c_k x^{k+\alpha}, \quad c_0 \neq 0, \text{ then } y' = \sum_{k=0}^{\infty} (k+\alpha) c_k x^{k+\alpha-1}, \text{ and}$$

* For the proof of this theorem see for example G. Birkoff and G.-C. Rota, *Ordinary differential equations*, p. 261, Blaisdell, Waltham, Mass., 1969.

$$y'' = \sum_{k=0}^{\infty}(k+\alpha)(k+\alpha-1)c_k x^{k+\alpha-2}$$

Substituting for y and its derivatives in the differential equation, we get

$$\sum_{k=0}^{\infty}[(k+\alpha)(k+\alpha-1)-4(k+\alpha)]c_k x^{k+\alpha-1}$$

$$-\sum_{k=0}^{\infty}(k+\alpha-2)c_k x^{k+\alpha} \equiv 0$$

To get the indicial equation, we equate the coefficient of the lowest power of x, which is $x^{\alpha-1}$, to zero, we obtain $\alpha(\alpha-5)c_0 = 0$, and since $c_0 \neq 0$, we have $\alpha(\alpha-5)=0$. Then $\alpha_1 = 5$ and $\alpha_2 = 0$. The roots are distinct and the difference between them is an integer (5). Now, equating the coefficients of $x^{k+\alpha}$ to zero, we obtain

$(k+\alpha+1)(k+\alpha-4)\,c_{k+1} = (k+\alpha-2)c_k$.

Rearranging, we get the recurrence relation

$$c_{k+1} = \frac{k+\alpha-2}{(k+\alpha+1)(k+\alpha-4)}c_k, \; k \geq 0.$$

For $\alpha = 5$, we have $c_{k+1} = \dfrac{k+5}{(k+6)(k+1)}c_k, \; k \geq 0.$

For various values of k, we have

$k=0$: $\quad c_1 = \dfrac{3}{6\cdot 1}c_0$

$k=1$: $\quad c_2 = \dfrac{4}{7\cdot 2}c_1 = \dfrac{3\cdot 4}{6\cdot 7\cdot 1\cdot 2}c_0$

$k=2$: $\quad c_3 = \dfrac{5}{8\cdot 3}c_2 = \dfrac{3\cdot 4\cdot 5}{6\cdot 7\cdot 8\cdot 1\cdot 2\cdot 3}c_0$

And for $k = n-1$: $\quad c_n = \dfrac{3\cdot 4\cdot 5}{(n+5)(n+4)(n+3)n!}c_0$

The first solution is $y_1 = c_0 \displaystyle\sum_{n=0}^{\infty} \frac{3\cdot 4\cdot 5}{(n+5)(n+4)(n+3)n!}x^{n+5}$.

Now, according to the previous theorem, the second solution will be of the form $y_2 = my_1 \ln x + \sum_{n=0}^{\infty} \beta_n x^n$.

Differentiating twice with respect to x, we obtain

$$y_2' = \frac{m}{x} y_1 + m y_1' \ln x + \sum_{n=1}^{\infty} n\,\beta_n x^{n-1} \quad \text{and}$$

$$y_2'' = -\frac{m}{x^2} y_1 + \frac{2m}{x} y_1' + m y_1'' \ln x + \sum_{n=2}^{\infty} n(n-1)\,\beta_n x^{n-2}.$$

Substituting these values into the differential equation, we obtain

$$m \ln x\,[x y_1'' - (4+x) y_1' + 2y_1] - m y_1\left(1+\frac{5}{x}\right) + 2m y_1'$$
$$+ \sum_{n=0}^{\infty} [(n+1)(n+2)\,\beta_{n+2} - (n+1)\beta_{n+1}]x^{n+1}$$
$$- \sum_{n=0}^{\infty} [4(n+1)\,\beta_{n+1} - 2\beta_n]x^n = 0$$

We can see that $\ln x$ is multiplied by $[x y_1'' - (4+x) y_1' + 2y_1]$ and since y_1 is a solution of the differential equation then $x y_1'' - (4+x) y_1' + 2y_1 = 0$, and the term containing $\ln x$ vanishes.

$$-m y_1\left(1+\frac{5}{x}\right) + 2m y_1' + \sum_{n=0}^{\infty} [(n+1)(n+2)\,\beta_{n+2} - (n+1)\beta_{n+1}]x^{n+1}$$
$$- \sum_{n=0}^{\infty} [4(n+1)\,\beta_{n+1} - 2\beta_n]x^n = 0$$

Now, substituting for y_1 (let $c_0 = 1$ for convenience), we obtain

$$-m\left(1+\frac{5}{x}\right) \sum_{n=0}^{\infty} \frac{3\cdot 4\cdot 5}{(n+5)(n+4)(n+3)n!} x^{n+5} + 2m \sum_{n=0}^{\infty} \frac{3\cdot 4\cdot 5}{(n+4)(n+3)n!} x^{n+4}$$
$$+ \sum_{n=0}^{\infty} [(n+1)(n+2)\,\beta_{n+2} - (n+1)\beta_{n+1}]x^{n+1} - \sum_{n=0}^{\infty} [4(n+1)\,\beta_{n+1} - 2\beta_n]x^n = 0$$

Equating the coefficients of various powers of x to zero, we get

$x^0 : -4\beta_1 + 2\beta_0 = 0$ $\qquad$ $\beta_1 = \frac{1}{2}\beta_0$

$x^1 : -6\beta_2 + \beta_1 = 0$ $\qquad$ $\beta_2 = \frac{1}{12}\beta_0$

$x^2 : -12\beta_3 + 2\beta_2 + 6\beta_3 - 2\beta_2 = 0$ $\qquad$ $\beta_3 = 0$

$x^3: -16\beta_4 + 2\beta_3 + 12\beta_4 - 3\beta_3 = 0 \qquad \beta_4 = 0$

$x^4: -20\beta_5 + 20\beta_5 + 2\beta_4 - 4\beta_4 + 10m = 0 \qquad \beta_5$ arbitrary, $m = 0$

In fact, β_5 represents the arbitrary constant for the first series solution and there is no need to proceed any further. The second solution is now $y_2 = \beta_0\left(1+\frac{1}{2}x+\frac{1}{12}x^2\right)$, and the general solution is

$$y = A\sum_{n=0}^{\infty}\frac{3\cdot 4\cdot 5}{(n+5)(n+4)(n+3)n!}x^{n+5} + B\left(1+\tfrac{1}{2}x+\tfrac{1}{12}x^2\right) \quad \square$$

Example 2: ***Logarithmic case***: Find the series solution for the following equation about $x = 0$: $xy'' - y = 0$.

Solution: The point $x = 0$ is a regular singular point of this differential equation. Assume a Frobenius series solution of the form

$$y = \sum_{k=0}^{\infty} c_k x^{k+\alpha}, \quad c_0 \neq 0, \text{ then } y' = \sum_{k=0}^{\infty}(k+\alpha)c_k x^{k+\alpha-1}, \text{ and}$$

$$y'' = \sum_{k=0}^{\infty}(k+\alpha)(k+\alpha-1)c_k x^{k+\alpha-2}$$

Substituting for y and its derivatives in the differential equation, we get

$$\sum_{k=0}^{\infty}(k+\alpha)(k+\alpha-1)c_k x^{k+\alpha-1} - \sum_{k=0}^{\infty} c_k x^{k+\alpha} \equiv 0$$

To get the indicial equation, we equate the coefficient of the lowest power of x, which is $x^{\alpha-1}$, to zero, we obtain $\alpha(\alpha-1)c_0 = 0$, and since $c_0 \neq 0$, we have $\alpha(\alpha-1) = 0$. Then $\alpha_1 = 1$ and $\alpha_2 = 0$. The roots are distinct and the difference between them is an integer ($\alpha_1 - \alpha_2 = 1$). Now, equating the coefficients of $x^{k+\alpha}$ to zero, we obtain $(k+\alpha+1)(k+\alpha)c_{k+1} = c_k$.

Rearranging, we get the recurrence relation

$$c_{k+1} = \frac{1}{(k+\alpha+1)(k+\alpha)}c_k, \quad k \geq 0.$$

<u>For $\alpha = 1$</u>, we have $c_{k+1} = \dfrac{1}{(k+1)(k+2)}c_k, \quad k \geq 0.$

For various values of k, we have

$k=0:\qquad c_1=\frac{1}{1\cdot 2}c_0$

$k=1:\qquad c_2=\frac{1}{2\cdot 3}c_1=\frac{1}{1\cdot 2\cdot 3\cdot 2}c_0$

$k=2:\qquad c_3=\frac{1}{3\cdot 4}c_2=\frac{1}{1\cdot 2\cdot 3\cdot 4\cdot 2\cdot 3}c_0$

And for $k=n-1$: $\quad c_n=\frac{1}{n!(n+1)!}c_0,\quad n\geq 1$

The first solution is $y_1=c_0\sum_{n=0}^{\infty}\frac{1}{n!(n+1)!}x^{n+1}$.

The second solution will be of the form $y_2=my_1\ln x+\sum_{n=0}^{\infty}\beta_n x^n$.

Differentiating twice with respect to x, we obtain

$y_2'=\frac{m}{x}y_1+my_1'\ln x+\sum_{n=1}^{\infty}n\,\beta_n x^{n-1}$ and

$y_2''=-\frac{m}{x^2}y_1+\frac{2m}{x}y_1'+my_1''\ln x+\sum_{n=2}^{\infty}n(n-1)\beta_n x^{n-2}$.

Substituting these values into the differential equation, we obtain

$$m\ln x\,[xy_1''-y_1]-\frac{m}{x}y_1+2my_1'+\sum_{n=0}^{\infty}(n+1)(n+2)\beta_{n+2}x^{n+1}-\sum_{n=0}^{\infty}\beta_n x^n=0$$

We can see that ln$x$ is multiplied by $[xy_1''-y_1]$ and since y_1 is a solution of the differential equation then $xy_1''-y_1=0$, and the term containing $\ln x$ vanishes.

$$-\frac{m}{x}y_1+2my_1'+\sum_{n=0}^{\infty}(n+1)(n+2)\beta_{n+2}x^{n+1}-\sum_{n=0}^{\infty}\beta_n x^n=0$$

Now, substituting for y_1 (let $c_0=1$ for convenience), we obtain

$$-m\sum_{n=0}^{\infty}\frac{1}{n!(n+1)!}x^n+2m\sum_{n=0}^{\infty}\frac{n+1}{n!(n+1)!}x^n+\sum_{n=0}^{\infty}(n+1)(n+2)\beta_{n+2}x^{n+1}-\sum_{n=0}^{\infty}\beta_n x^n=0$$

Equating the coefficient of x^0 to zero, we obtain

$$-m + 2m - \beta_0 = 0 \quad \text{or} \quad m = \beta_0$$

Equating the coefficients of x^n to zero, we get

$$\frac{2m(n+1)}{n!(n+1)!} - \frac{m}{n!(n+1)!} + n(n+1)\beta_{n+1} - \beta_n = 0$$

Rearranging, the recurrence relation for the coefficients β_n is

$$\beta_{n+1} = \frac{1}{n(n+1)}\left[\beta_n - \frac{\beta_0(2n+1)}{n!(n+1)!}\right], \quad n \geq 1,$$

and for various values of *k*, we have

$n = 1$: $\quad \beta_2 = \frac{1}{2}\beta_1 - \frac{3}{4}\beta_0$

$n = 2$: $\quad \beta_3 = \frac{1}{12}\beta_1 - \frac{7}{36}\beta_0$

$n = 3$: $\quad \beta_4 = \frac{1}{144}\beta_1 - \frac{35}{1728}\beta_0$

The second solution is

$$y_2 = \beta_0 y_1 \ln x + \beta_0 + \beta_1 x + \left(\tfrac{1}{2}\beta_1 - \tfrac{3}{4}\beta_0\right)x^2 + \left(\tfrac{1}{12}\beta_1 - \tfrac{7}{36}\beta_0\right)x^3 + \left(\tfrac{1}{144}\beta_1 - \tfrac{35}{1728}\beta_0\right)x^4 + \cdots$$

or

$$y_2 = \beta_0 y_1 \ln x + \beta_0\left(1 - \tfrac{3}{4}x^2 - \tfrac{7}{36}x^3 - \tfrac{35}{1728}x^4 - \cdots\right) + \beta_1\left(x + \tfrac{1}{2}x^2 + \tfrac{1}{12}x^3 + \tfrac{1}{144}x^4 + \cdots\right)$$

We can recognize that the expression multiplied by β_1 is nothing but the first solution. Then we can drop it from the second solution to obtain

$$y_2 = \beta_0\left[y_1 \ln x + \left(1 - \tfrac{3}{4}x^2 - \tfrac{7}{36}x^3 - \tfrac{35}{1728}x^4 - \cdots\right)\right],$$

and the general solution is

$$y = A\sum_{n=0}^{\infty}\frac{x^{n+1}}{n!(n+1)!} + B\left[\ln x \sum_{n=0}^{\infty}\frac{x^{n+1}}{n!(n+1)!} + \left(1 - \tfrac{3}{4}x^2 - \tfrac{7}{36}x^3 - \tfrac{35}{1728}x^4 - \cdots\right)\right] \quad \square$$

Exercise 5.5

Obtain series solutions for the following equation about $x=0$.

1. $xy''+2x(x-2)y'+2(2-3x)y=0$

Ans: $y=c_0\left(x-2x^2+2x^3\right)+c_3\sum_{n=3}^{\infty}\frac{6(-2)^{n-3}x^{n+1}}{n!}$

2. $xy''-(3+x)y'+2y=0$

Ans: $y=c_0\left(1+\frac{2}{3}x+\frac{1}{6}x^2\right)+c_4\sum_{n=4}^{\infty}\frac{24(n-3)x^n}{n!}$

3. $x(1-x)y''-3y'+2y=0$

Ans: $y=c_0\left(1+\frac{2}{3}x+\frac{1}{3}x^2\right)+c_4\sum_{n=4}^{\infty}(n-3)x^n$

4. $xy''+(4+3x)y'+3y=0$

Ans: $y=c_0\left(x^{-3}-3x^{-2}+\frac{9}{2}x^{-1}\right)+6c_3\sum_{n=3}^{\infty}\frac{(-3)^{n-3}x^{n-3}}{n!}$

5. $x(x+3)y''-9y'-6y=0$

Ans: $y=c_0\left(1-\frac{2}{3}x+\frac{1}{3}x^2-\frac{4}{27}x^3\right)+c_4\sum_{n=4}^{\infty}\frac{(-1)^n(n+1)x^n}{5\cdot 3^{n-4}}$

6. $x^2y''+x^2y'-2y=0$

Ans: $y=c_0\left(x^{-1}-\frac{1}{2}\right)+6c_3\sum_{n=3}^{\infty}\frac{(-1)^{n+1}(n-2)x^{n-1}}{n!}$

7. $xy''+(x^3-1)y'+x^2y=0$

Ans: $y=c_0\sum_{n=0}^{\infty}\frac{(-1)^n x^{3n}}{3^n n!}+c_2\left[x^2+\sum_{n=1}^{\infty}\frac{(-1)^n x^{3n+2}}{5\cdot 8\cdot 11\cdot\dots\cdot(3n+2)}\right]$

8. $x^2y''+x(x-2)y'+(x^2+2)y=0$

Ans: $y_1=x^2-x^3+\frac{1}{3}x^4-\frac{1}{36}x^5-\frac{7}{720}x^6+\dots$

$y_2=y_1\ln x-x+\frac{3}{2}x^3-\frac{31}{36}x^4+\frac{65}{432}x^5+\frac{61}{4320}x^6+\dots$

9. $xy'' + (x-1)y' - 2y = 0$

Ans: $y_1 = x^2,\ y_2 = -y_1 \ln x + 1 - 2x + 2x^2 - 2\sum_{n=3}^{\infty} \frac{(-1)^n x^n}{n!(n-2)}$

10. $x^2 y'' - 3xy' + (x^3 - 5)y = 0$ **Ans:** $y_1 = 2\sum_{n=0}^{\infty} \frac{(-1)^n x^n}{3^{2n} n!(n+1)!}$

$$y_2 = -\tfrac{1}{54} y_1 \ln x + x^{-1} + \tfrac{1}{9} x^2 + \tfrac{1}{324} x^5 - \sum_{n=3}^{\infty} \frac{(-1)^n x^{3n-1}}{3^{2n} n!(n-2)!}\left[1 + \frac{1}{n-1} + \frac{1}{n} - 2\left(1 + \frac{1}{2} + \frac{1}{3} + \cdots + \frac{1}{n}\right)\right]$$

11. $x(1-x)y'' - 4xy' - 2y = 0$

Ans: $y_1 = \sum_{n=0}^{\infty} n(n+1)x^n,\ y_2 = y_1 \ln x + \sum_{n=0}^{\infty} (1+n-n^2)x^n$

12. $x(1-x)y'' + 2(1-x)y' + 2y = 0$

Ans: $y_1 = 2x - 2,\ y_2 = y_1 \ln x + x^{-1} + 1 - 5x + \sum_{n=3}^{\infty} \frac{2x^{n-1}}{(n-1)(n-2)}$

13. $xy'' + (1-x)y' + 3y = 0$

Ans: $y_2 = y_1 \ln x + 7x - \frac{23}{4} x^2 + \frac{11}{12} x^3 - 6\sum_{n=4}^{\infty} \frac{x^n}{n!n(n-1)(n-2)(n-3)}$

14. $x^2 y'' + x(1-x)y' + 2xy = 0$ **Ans:**

$y_1 = 2 - 4x + x^2, y_2 = y_1 \ln x + 10x - \frac{9}{2} x^2 + \frac{1}{9} x^3 + \frac{1}{144} x^4 + \frac{1}{1800} x^5 + \cdots$

15. $x(1-x)y'' + 4y' + 2y = 0$

Ans: $y = a\left\{1 - \frac{x}{2} + \frac{x^2}{10} + \cdots\right\} + \frac{b}{x^3}\left\{1 - 5x + 10x^2 + \cdots\right\}$

16. $xy'' + y' + x^2 y = 0$

Ans: $y = (a + b\ln x)\left\{1 - \frac{x^3}{3^2} + \frac{x^6}{3^4(2!)^2} + \cdots\right\} + 2b\left\{\frac{x^3}{3^3} - \frac{\left(1+\frac{1}{2}\right)}{3^5(2!)^2} x^6 + \cdots\right\}$

5.5. Solutions for Large Values of x

Sometimes it is desirable to obtain series solutions for differential equations as x, the independent variable, approaches infinity. This case is often called the asymptotic behavior of the differential equation. In this case, we use the transformation $x = 1/u$.

The behavior at the point of infinity is obtained by studying the behavior of the transformed equation at $u = 0$. If $u = 0$ is an ordinary point of the transformed equation, then the point at infinity for the original equation is also ordinary. The same apply for regular and irregular singular points.

***Example* 1**: Determine the nature of the point at infinity for the differential equation $2x(x+1)\dfrac{d^2y}{dx^2} + (5x-3)\dfrac{dy}{dx} + y = 0$, then obtain a series solution for large values of x.

Solution: Let $x = \dfrac{1}{u}$, then we have $\dfrac{dy}{dx} = \dfrac{dy}{du}\cdot\dfrac{du}{dx} = -u^2\dfrac{dy}{du}$

and $\dfrac{d^2y}{dx^2} = \dfrac{d}{dx}\left(-u^2\dfrac{dy}{du}\right) = u^4\dfrac{d^2y}{du^2} + 2u^3\dfrac{dy}{du}$

Substituting these expressions into the differential equation and rearranging, we obtain

$$2u^2(u+1)\frac{d^2y}{du^2} + u(7u-1)\frac{dy}{du} + y = 0$$

Clearly $u = 0$ is a regular singular of this point equation. Then, the point at infinity for the original equation is also a regular singular point.

The solution of the transformed equation is found to be

$$y = A\sum_{n=0}^{\infty}(-1)^n(2n+3)(2n+5)u^{n+1} + B\sum_{n=0}^{\infty}(-1)^n(n+1)(n+2)u^{n+1/2}$$

Then the solution of the original equation for large values of x is

$$y = A\sum_{n=0}^{\infty}(-1)^n(2n+3)(2n+5)x^{-n-1} + B\sum_{n=0}^{\infty}(-1)^n(n+1)(n+2)x^{-n-\frac{1}{2}}$$

***Example* 2**: Determine the nature of the point at infinity for the differential equation $(1-x^2)y'' - 2x\,y' + 6y = 0$, then obtain a series solution for large values of x.

Solution: Let $x = \dfrac{1}{u}$, then we have $\dfrac{dy}{dx} = \dfrac{dy}{du}\cdot\dfrac{du}{dx} = -u^2\dfrac{dy}{du}$

and $\frac{d^2y}{dx^2} = \frac{d}{dx}\left(-u^2\frac{dy}{du}\right) = u^4\frac{d^2y}{du^2} + 2u^3\frac{dy}{du}$

Substituting these expressions in the differential equation and rearranging, we obtain

$$u^2(u^2-1)\frac{d^2y}{du^2} + 2u^3\frac{dy}{du} + 6y = 0.$$

Here, $P(u) = \frac{2u}{u^2-1}$ and $Q(u) = \frac{6}{u^2(u^2-1)}$. Clearly $u = 0$ is a regular point of the differential equation, and the point at infinity is also a regular point for the original differential equation. Assume that the solution is of the form

$u = \sum_{k=0}^{\infty} c_k u^{k+\alpha}$, then $u' = \sum_{k=0}^{\infty} (k+\alpha)c_k u^{k+\alpha-1}$ and

$$u'' = \sum_{k=0}^{\infty} (k+\alpha)(k+\alpha-1)c_k u^{k+\alpha-2}.$$

Substituting these vales in the differential equation, we get

$$u^2(u^2-1)\sum_{k=0}^{\infty} (k+\alpha)(k+\alpha-1)c_k u^{k+\alpha-2}$$

$$+ 2u^3\sum_{k=0}^{\infty} (k+\alpha)c_k u^{k+\alpha-1} + 6\sum_{k=0}^{\infty} c_k u^{k+\alpha} = 0$$

Equating the coefficient of the lowest power of u, (u^{α}), to zero, to obtain $-\alpha(\alpha-1)c_0 + 6c_0 = 0$. Since $c_0 \neq 0$, the ***indicial equation*** is $\alpha^2 - \alpha - 6 = 0$. The roots are $\alpha = 3, -2$.

Equating the coefficients of $u^{\alpha+1}$ to zero, we obtain
$(\alpha+1)\alpha c_1 + 6c_1 = 0$. Since $\alpha = 3, -2$, then c_1 must be zero.

Equating the coefficients of $u^{k+\alpha}$ to zero, we obtain

$$(k+\alpha-2)(k+\alpha-3)c_{k-2} - (k+\alpha)(k+\alpha-1)\,c_k$$

$$+2(k+\alpha-2)c_{k-2} + 6c_k = 0$$

Rearranging, we obtain the recurrence relation for the coefficients as

$$\boxed{c_k = \frac{(k+\alpha-2)(k+\alpha-1)}{(k+\alpha-3)(k+\alpha+2)}c_{k-2}, \quad k \geq 2}.$$

For various values of k, we have

$$k=2:\quad c_2=\frac{\alpha(\alpha+1)}{(\alpha-1)(\alpha+4)}c_0$$

$$k=4:\quad c_4=\frac{\alpha(\alpha+2)(\alpha+3)}{(\alpha-1)(\alpha+4)(\alpha+6)}c_0$$

The solution in terms of α is

$$y=c_0u^{\alpha}\left\{1+\frac{\alpha(\alpha+1)}{(\alpha-1)(\alpha+4)}u^2+\frac{\alpha(\alpha+2)(\alpha+3)}{(\alpha-1)(\alpha+4)(\alpha+6)}u^4+\cdots\right\}$$

For $\alpha=3$, the first solution is

$$y_1=au^3\left\{1+\frac{3\cdot4\cdot u^2}{2\cdot7}+\frac{3\cdot5\cdot6\cdot u^4}{2\cdot7\cdot9}+\cdots\right\}.$$

For $\alpha=-2$, The second solution is $y_2=\dfrac{b}{u^2}\left\{1-\dfrac{u^2}{3}\right\}$.

The general solution will be

$$y=au^3\left\{1+\frac{3\cdot4\cdot u^2}{2\cdot7}+\frac{3\cdot5\cdot6\cdot u^4}{2\cdot7\cdot9}+\cdots\right\}+\frac{b}{u^2}\left\{1-\frac{u^2}{3}\right\}.$$

Replacing u by $1/x$, the general solution of the original differential equation is

$$y=\frac{a}{x^3}\left\{1+\frac{3\cdot4}{2\cdot7\cdot x^2}+\frac{3\cdot5\cdot6}{2\cdot7\cdot9\cdot x^4}+\cdots\right\}+b\left\{x^2-\frac{1}{3}\right\}.$$ □

Exercise 5.6

Determine the nature of the point at infinity for the following differential equations, then find the series solutions for large values of x:

1. $2x^3y''+x(3x-1)y'-2y=0$

   Ans: $y=a\left(1+\dfrac{2}{x}+\dfrac{1}{3x^2}\right)+b\,x^{-1/2}\displaystyle\sum_{n=0}^{\infty}\frac{(-1)^n3x^{-n}}{2^n\,n!(2n-3)(2n-1)(2n+1)}$

2. $3x^3y''+x(4x+1)y'-2y=0$

   Ans: $y=a\displaystyle\sum_{n=0}^{\infty}\frac{(3n+4)}{4\cdot3^n\,n!}x^{-n-1/3}+b\left[1+\sum_{n=1}^{\infty}\frac{(n+1)x^{-n}}{2\cdot5\cdot8\cdot\cdots\cdot(3n-1)}\right]$

3. $(1-x^2)y''-2xy'+2y=0$

   Ans: $y=ax^{-1}+bx^{-2}\left(1+\dfrac{3}{5x^2}+\dfrac{3}{7x^4}+\cdots\right)$

www.ingramcontent.com/pod-product-compliance
Lightning Source LLC
Chambersburg PA
CBHW032011170526
45157CB00002B/650
* 9 7 8 1 4 1 1 6 3 9 2 0 1 *